# *Agri-Culture*

This is for Gill, Freya and Theo
and for Mum and Dad for the pictures and advice
and time too for a mention for Chris, Dawn, Fern, Joss,
Tom and William; Annabel, Mike, Jasper and Felix;
Pat and John

# Agri-Culture
## Reconnecting People, Land and Nature

### Jules Pretty

Earthscan Publications Limited
London • Sterling, VA

First published by Earthscan in the UK and USA in 2002

Reprinted 2003, 2007

Copyright © Jules Pretty, 2002

ISBN: 978 1 85383 925 2 paperback
      978 1 85383 920 7 hardback

Typesetting by JS Typesetting Ltd, Porthcawl, Mid Glamorgan
Printed and bound in the UK by Biddles, King's Lynn
Cover design by Adrian Senior
Cover illustration by John R Pretty

For a full list of publications please contact:

**Earthscan**
8–12 Camden High Street
London, NW1 0JH, UK
Tel: +44 (0)20 7387 8558
Fax: +44 (0)20 7387 8998
Email: earthinfo@earthscan.co.uk
Web: **www.earthscan.co.uk**

22883 Quicksilver Drive, Sterling, VA 20166-2012, USA

Earthscan is an imprint of James and James (Science Publishers) Ltd and
publishes in association with the International Institute for Environment and
Development

A catalogue record for this book is available from the British Library

Library of Congress Cataloging-in-Publication Data

Pretty, Jules N.
     Agri-culture : reconnecting people, land and nature / by Jules Pretty.
         p. cm.
     Includes bibliographical references (p. ).
     ISBN 1-85383-925-6 (pbk.) – ISBN 1-85383-920-5 (hardback)
         1. Sustainable agriculture. 2. Agriculture–Social aspects.
    3. Agricultural productivity. I. Title.
    S494.5.S86 P73 2002
    333.76′16–dc21

                                    2002009735

# Contents

# Acknowledgements

I am very grateful to the many people who provided invaluable critical comment on earlier drafts of this book and advice on related material. I would like to make special mention of Ted Benton, James Morison and Norman Uphoff, who went well beyond the call of duty, together with Curtis Absher, Jacqui Ashby, Richard Aylard, Neil Baker, Andy Ball, David Beckingsale, Phil Bradley, Lynda Brown, Susanne and Simon Campbell-Jones, Robert Chambers, Nigel Cooper, Ed Cross, Jan Deane, John Devavaram, Amadou Diop, Tom Dobbs, David Favis-Mortlock, Bruce Frank, Phil and Susie Grice, Julia Guivant, Jilly Hall, John Hall, Brian Halweil, Hal Hamilton, Justin Hardy, Sue Heisswolf, Rachel Hine, Ian Hutchcroft, J K Kiara, John Landers, Tim Lang, Howard Lee, David Lort-Phillips, Simon Lyster, Joe Morris, Phil Mullineaux, Eri Nakajima, Hiltrud Nieberg, Kevin Niemeyer, David Orr, Roberto Peiretti, Michel Pimbert, Tim O'Riordan, Mark Ritchie, Colin Samson, Pedro Sanchez, Sara Scherr, Dorothy and Walter Schwarz, Mardie Townsend, Hugh Ward, Drennan Watson, Jane Weissman, Mark Winne, and Vo-Tong Xuan. The views represented here are, of course, all my own responsibility, as are the mistakes.

As ever, I am grateful to Jonathan Sinclair Wilson, and to Akan Leander and all at Earthscan, for valuable support and advice, and to Marie Chan for administrative support.

The illustrations in this book, including the cover painting from northern Nigeria, are all painted by John R Pretty who is a professional painter specializing in marine and landscape watercolours of East Anglia and France. He is a member of the East Anglian Group of Marine Artists with whom he exhibits at the Mall Gallery in London, and he has been guest of honour at various Salons in France, being the recipient of ten medals for watercolour. He has exhibited at the Royal Society of Marine Artists and the National Society of Painters, Sculptors and Printmakers, and, in France, at the Gallerie du Museé, Les Baux-de-Provence, and the Salon of the Societé International des Beaux Arts Paris and in Salons at Parçay

Mesley, Le Poinçonnet, Tinchebray, Argenton-sur-Creuse and Mongermont. Annually, he exhibits at the Salon du Val d'Or of the Association Internationale Plastica Latina, with the Mérite Artistique Européen France, at the Salon in St Florent-sur-Cher, where he is a member, and at the Salon des Amis de Montemartre, Issoudun. His work is also in collections in Europe, America, New Zealand and South Africa, and can be seen at www.johnrpretty.co.uk.

# Glossary of Specialist Terms

Terms that appear in the glossary are italicized on their first mention in each chapter of the text.

| | |
|---|---|
| *Adat* | Indigenous cultural knowledge and rules system of Indonesia and Malaysia |
| Agroecology | Ecological relationships in agricultural systems |
| Antimicrobial (also antibiotic) | Substance toxic to microbes and administered to humans and livestock |
| Apomixis | The production of exact clones of the mother plant through asexual reproduction |
| Autopoiēsis | Self-organizing and self-made character of living systems |
| Bioregionalism | Integration of human activities within ecological limits |
| Biotechnology | Molecular changes to living or non-living things, involving the transfer of DNA from one organism to another, allowing the recipient to express new characteristics |
| Carbon sequestration | Capturing or locking up of carbon dioxide from the atmosphere in sinks (eg soils, trees) |
| CBD | Convention on Biological Diversity |
| Cloning | The production of individuals with an exact copy of the DNA of another organism |
| Common property resource | Resources used in common by a defined group of people, usually with locally developed rules for their use |
| Cryptosporidium | Pathogenic organism arising from domestic livestock and wild animals, and pollutant of water |

| | |
|---|---|
| CSA | Community-supported agriculture farm |
| DFID | Department for International Development (UK) |
| Diverscape | An ecologically and/or socially diverse landscape |
| Enclosure (also inclosure) | The changing of large commonly managed open fields in Europe to produce smaller individually owned fields enclosed by hedges or other boundaries |
| ESA | Environmentally sensitive area |
| Eutrophication | Nutrient enrichment of water that leads to excessive algal growth, disruption of whole food webs and, in the worst cases, complete eradication of all life through deoxygenation |
| Extensionist | Agricultural professional in regular contact with both farmers and researchers |
| Externality | Any action that affects the welfare of, or opportunities available to, an individual or group without direct payment or compensation, and may be positive or negative |
| FAO | Food and Agriculture Organization (UN) |
| Fen | Grassland by river, occasionally flooded |
| Food security | Access to sufficient food of appropriate diversity for a healthy diet |
| Foodshed | Self-reliant, locally or regionally based food systems comprised of diversified farms using sustainable practices to which consumers are linked by the bonds of community as well as economy |
| GM | Genetically modified |
| Hectare | Measure of area equivalent to 2.47 acres |
| HFS | Hartford Food System (US) |
| ICIPE | International Centre for Insect Physiology and Ecology |
| IFPRI | International Food Policy Research Institute |
| Integrated pest management | The use of a variety of methods and approaches to manage pests and diseases, usually minimizing or even eliminating pesticide use |

| | |
|---|---|
| Megajoule (MJ) | Measure of energy, with one MJ equal to 4.2 kilocalories (kcals) |
| Microfinance | System of savings and credit run by local groups, often with some external matched funds |
| Modernist | Single-coded, inflexible and monocultural systems |
| Monoscape | A landscape without ecological or social diversity |
| Multifunctional | Agricultural systems with many side effects in addition to food production |
| Nutrition transition | Effect of increasing urbanization on people's adoption of new diets, resulting, in particular, in consumption of more meat and fewer traditional cereals |
| Public goods | Goods or services which when consumed by a group member cannot be withheld from other members of the group, or when consumed can still be consumed by other members of the group |
| Rhizobia | Soil microflora aiding nitrogen fixation |
| Salinization | Process of salt build-up in soils, arising from over-irrigation in wetlands or from removal of vegetation in drylands |
| Satochi | Landscapes in Japan marked by great diversity in the relationships between humans and nature |
| Semiochemical | Aromatic chemicals given off by plants that attract or repel insect pests, predators or parasites |
| SRI | System of Rice Intensification |
| Stakeholder | Person or group with a particular stake or interest in an activity or organization |
| Subak | Irrigation cooperatives in Bali |
| Transgene | Gene transferred from one organism to another not by conventional breeding |
| Watershed | Topographic area draining to a single point |
| WHO | World Health Organization |
| Zero-tillage | Farming without inversion ploughing in which seeds are direct-drilled and the soil surface remains permanently covered |

# Preface to a Revolution

Something is wrong with our agricultural and food systems. Despite great progress in increasing productivity during the last century, hundreds of millions of people remain hungry and malnourished. Further hundreds of millions eat too much, or the wrong sorts of food, and it is making them ill. The health of the environment suffers too, as degradation seems to accompany many of the agricultural systems we have evolved in recent years. Can nothing be done, or is it time for the expansion of another sort of agriculture, founded more on ecological principles and in harmony with people, their societies and cultures? This is not a new idea, as many have struggled in the past to come up with both sustainable and productive farm systems, and have had some success. What is novel, though, is that these systems are now beginning to spread to many new places, and are reaching a scale large enough to make a difference to the lives of millions of people.

My intention in writing this book is to help to popularize this complex and rather hidden area of human endeavour. I live and work in the picturesque landscape of the Suffolk and Essex borders of eastern England, a region of small fields, ancient hedgerows, lazy rivers and Tudor wool towns. I spent my early years growing up amongst the sands and savannahs of the Sahara's southern edge, landscapes dotted with baobab and acacia, and teeming with wildlife. In my time, I have had the fortune to meet and work with inspiring people in many communities in both developing and industrialized countries. Most have been swimming against a prevailing tide of opinion, often exposing themselves to ridicule or even opprobrium. In writing this book, I want to tell some of their stories, about how individuals and groups have chosen routes to transformation, and how they have succeeded in changing both communities and landscapes.

I also want to present evidence to support the contention that industrialized agricultural systems as currently configured are flawed, despite their great progress in increasing food productivity, and that alternative systems can be efficient and equitable. My intention is to bring these ideas to a wider audience, because food matters to us all. As consumers, we buy it every week, even every day, and the choices we make send strong signals about the systems of agricultural production that we prefer. We may not realize that these messages are being sent, but they are. Our daily consumption of food fundamentally affects the landscapes, communities and environments from which it originates.

In the earliest surviving texts on European farming, agriculture was interpreted as two connected things, *agri* and *cultura*, and food was seen as a vital part of the cultures and communities that produced it. Today, however, our experience with industrial farming dominates, with food now seen simply as a commodity, and farming often organized along factory lines. The questions I would like to ask are these. Can we put the culture back into agri-culture without compromising the need to produce enough food? Can we create sustainable systems of farming that are efficient and fair and founded on a detailed understanding of the benefits of agro-ecology and people's capacity to cooperate?

As we advance into the early years of the 21st century, it seems to me that we have some critical choices. Humans have been farming for some 600 generations, and for most of that time the production and consumption of food has been intimately connected to cultural and social systems. Foods have a special significance and meaning, as do the fields, grasslands, forests, rivers and seas. Yet, over just the last two or three generations, we have developed hugely successful agricultural systems based on industrial principles. They certainly produce more food per hectare and per worker than ever before, but only look efficient if we ignore the harmful side effects — the loss of soils, the damage to biodiversity, the pollution of water, the harm to human health.

Over these 12,000 years of agriculture, there have been long periods of stability, punctuated by short bursts of rapid change. These resulted in fundamental shifts in the way people thought and acted. I believe we are at another such junction. A sustainable agriculture making the best of nature and people's knowledge and collective capacities has been showing increasingly good promise. But it has been a quiet revolution because many accord it little credence. It is also silent because those in the vanguard are often the poorest and marginalized, whose voices are rarely heard in the grand scheme of things. No one can exactly say where this revolution might lead us. Neither do we know whether sustainable models of production would be appropriate for all farmers worldwide. But what I do know is that the principles apply widely. Once these come to be accepted, then it will be the ingenuity of local people that shapes these new methods of producing food to their own particular circumstances.

We know that most transitions involve trade-offs. A gain in one area is accompanied by a loss elsewhere. A road built to increase access to markets helps remote communities, but also allows illegal loggers to remove valuable trees more easily. A farm that eschews the use of pesticides benefits biodiversity, but may produce less food. New agroecological methods may mean more labour is required, putting an additional burden on women. But these trade-offs need not always be serious. If we listen carefully, and observe the improvements already being made by communities across the world, we find that it is possible to produce more food whilst protecting and improving nature. It is possible to have diversity in both human and natural systems without undermining economic efficiency.

This book draws on many stories of successful transformation. Sadly, I cannot do them full justice; as a result, they are inevitably partial. Nor is there the space to provide a careful consideration of all possible drawbacks or contradictions. I do not want to give the impression that just because some communities and societies are designated as 'traditional' or 'indigenous' they are always somehow virtuous, both in their relations with nature and with each other. The actions of some communities have led to ecological destruction. The norms of others have seen socially divisive and inequitable relations persist for centuries. Nonetheless, my intention here is to show what is possible, on both the ecological and social fronts, and not necessarily to imply that each and every case is perfect. This is also not a book where readers will find substantial evidence and analysis. There are no tables or figures in the main text, though the endnotes do contain much primary data. I am convinced, however, that the stories are based on sound methods and trustworthy evidence, and that they represent a significance beyond the specificities of their own circumstances.

I anticipate criticism from those who disbelieve that such progress can be made with agroecological approaches. I also do not want to reject all recent achievements in agriculture by presenting a doctrinaire alternative. Real progress can only come from a synthesis of the best of the past, eliminating practices that cause damage to environments and human health, and using the best of knowledge and technologies available to us today.

This sustainable agriculture revolution is now helping to bring forth a new world. But it is not likely to happen easily. Many agricultural policies are unhelpful. Many institutions do not listen to the voices of local people, particularly if they are poor or remote. Many companies still think that maximizing profit at a cost to the environment represents responsible behaviour. However, changing national or local policies is only one step. Governments may wish for certain things; but having the political will does not necessarily guarantee a desired outcome. Structural distortions in economies, self-interest, unequal trading relations, corruption, debt burdens, profit-maximization, environmental degradation, and war and conflict all reduce the likelihood of achieving the systemic change required to nurture this emerging revolution.

But we must not let these deep problems stop us from trying. Things change when enough people want them to. The time is surely right to speak loudly and, with a collective will, seek any innovations that will help overcome these problems. This book aims to take readers on a short journey through some of the communities and farms of both developing and industrialized countries where progress is being made. I hope you will agree that these stories of success deserve careful consideration and some celebration.

Chapter I sets the scene by showing that landscapes, and their attendant agricultural and food systems, are a common heritage to us all. In the pursuit of improved agricultural productivity, we have, nevertheless, allowed ourselves to become disconnected from nature, and so tend not to notice when it is damaged or taken away. For all our human history, we have been shaped by nature, while shaping it in return. But in our industrial age, we are losing the stories, memories and language about land and nature. These disconnections matter, for the way we think about nature and wildernesses fundamentally affects what we do in our agricultural and food systems.

Chapter 2 focuses on the darker side of the landscape, showing how the poor and powerless are commonly excluded from the very resources upon which they rely for their livelihoods. Modern dispossessions have extended such actions both in the name of economic growth and in the name of nature conservation. Strictly protected areas that are designed to protect biodiversity simply disconnect us once again from the nature we value and need. At the same time, modern agriculture has created monoscapes in order to enhance efficiency, and the poorest have lost out again. Repossession and regeneration of diverse and culturally important landscapes is an urgent task.

Chapter 3 takes a deliberately narrow economic perspective on the real costs and benefits of agricultural systems. The real price of food should incorporate the substantial *externalities*, or negative side effects, that must be paid for in terms of the harm to the environment and human health. Food appears cheap because these costs are difficult to identify and measure. Allocating monetary values to nature's goods and services is only one part of the picture; but it does tell us something of the comparative value of sustainable and non-sustainable systems, as well as indicate the kind of directions national policies should be taking. To date, the fine words of governments have only very rarely been translated into coherent and effective policies that support sustainable systems of food production. Chapter 4 shows how food poverty can be eliminated with more sustainable agriculture. We know that modern technologies and fossil-fuel derived inputs can increase agricultural productivity. However, anything that costs money inevitably puts it out of the reach of the poorest households and countries. Sustainable agriculture seeks to make the best use of nature's goods and services, of the knowledge and skills of farmers, and of people's collective capacity to work together to solve common management problems. Such systems relate to improving soil health, increasing water efficiency and reducing dependency on pesticides. When put together, the emergent systems are both diverse and productive. There are, of course, many threats, which may come to undermine much of the remarkable progress.

Chapter 5 focuses on the need to reconnect whole food systems. Industrialized countries have celebrated their agricultural systems' production of commodities; yet family farms have disappeared as rapidly as rural biodiversity. At the same time, farmers themselves have received a progressively smaller proportion of what consumers spend on food. Putting sustainable systems of production in touch with consumers within bioregions or *foodsheds* offers opportunities to recreate some of the connections. Farmers' markets, community-supported agriculture, box schemes and farmers groups are all helping to demonstrate what is possible. None of these alone will provoke systemic change, though regional policies and movements are helping to create the right conditions.

Chapter 6 addresses the genetic controversy. It is impossible to write of agricultural transformation without also assessing *biotechnology* and genetic modification. Who produces agricultural technologies, how they can be made available to the poor, and whether they will have adverse environmental effects, are all important questions we should ask regarding the many different types of genetic modification and different generations of application. The answers will tell us whether these new ideas can make a difference. We must, therefore, treat biotechnologies on a case-by-case basis, carefully assessing the potential benefits as well as the environmental and health risks. It is likely that biotechnology will make some contributions to the sustainability of agricultural systems; but developing the research systems, institutions and policies to make them pro-poor will be much more difficult.

Chapter 7 centres on the need to develop social learning systems to increase ecological literacy. Our knowledge of nature and the land usually accrues slowly over time, and cannot easily be transferred. If an agriculture dependent upon detailed ecological understanding is to emerge, then social learning and participatory systems are a necessary prerequisite. These develop relations of trust, reciprocal mechanisms, common rules and norms, and new forms of connectedness institutionalized in social groups. New commons are now being created for the collective management of *watersheds*, water, *microfinance*, forests and pests. These collective systems, involving the emergence of some 400,000 groups over just a decade, can also provoke significant personal changes. No advance towards sustainability can occur without us crossing the internal frontiers, too.

Chapter 8 focuses on a select number of cases and individuals who have crossed the internal frontiers and then caused large-scale external transformations. Our old thinking has failed the rest of nature, and is in danger of failing us again. Could we help to make a difference if we changed the way we think and act? Can we, as Aldo Leopold suggested, think like the mountain and the wolf? Heroic change is possible, yet we also need to expand from the parochiality of these cases. Everyone is in favour of sustainability, yet few seriously go beyond the fine words. There really is no alternative to the radical reform of national agricultural, rural and food policies, and institutions. The need is urgent, and this is not the time to hesitate. The time has come for this next agricultural revolution.

Jules Pretty
Professor of Environment and Society
University of Essex
December 2001

*Chapter 1*

# Landscapes Lost and Found

## This Common Heritage

In a bend of the river stands an ancient, open meadow. These 30 hectares of the *Fen*, as it is known hereabouts, are a relic. For 600 years, the flint church tower has gazed through village trees upon an ever-changing agricultural landscape. This common, though, has survived intact. It is parcelled into 180 'fennages', or rights to graze cattle, and so is in common ownership. When the harsh easterly winds drive down from Scandinavia, the grass crunches underfoot, and the pasture hollows are thick with ice. On a summer's day, you walk the same route past carpets of yellow buttercups, or divert past an enclosed hay meadow dotted with purple bee orchids. In autumn, after a few days of rain, the river floods and spills upon the pastures, lighting the landscape with the colour of the sky. In the long evenings, bats flit through clouds of insects, and owls hoot in search of scurrying prey. Splashes from the river remind us of the mysterious lives of otters. This Fen is different from the surrounding farmland, and it has been this way for centuries.

Other things are important about this common meadow. It links local people with nature, and as it is used and valued as a common, so it connects rights owners and users with one another. In recent years, though, both of these types of connection have been widely neglected and consequently eroded – to our loss, and to the loss of nature at large. As food has become a commodity, most of us no longer feel a link to the place of production and its associated culture. Yet agricultural and food systems, with their associated nature and landscapes, are a common heritage and thus, also, a form of common property. They are shaped by us all, and so in some way are part of us all, too. Landscapes across the world have been created through our interactions with nature. They have emerged through history, and have become deeply embedded in our cultures and consciousness. From the rural idylls of England to the diverse *satochi* of Japan, from the terraced rice fields and tree-vegetable gardens of Asia to the savannahs of Africa and forests of the Amazon, they have given collective meaning to whole societies, imparting a sense of permanence and stability. They are places that local people know, where they feel comfortable, where they belong.

When we feel that we have ownership in something, even if technically and legally we do not, or that our livelihood depends upon it, then we care. If we care, we watch, we appreciate, we are vigilant against threats. But when we know less, or have forgotten, we do not care. Then it is easier for the powerful to appropriate these common goods and so destroy them in pursuit of their own economic gain. For more than 100 centuries, cultivators have tamed the wilderness – controlling and managing nature, mostly with a sensitive touch. But all has changed in the last half per cent of that time. The rapid modernization of landscapes in both developing and industrialized countries has broken many of our natural links with land and food, and so undermined a sense of ownership, an inclination to care, and a desire to take action for the collective good.

Sometimes the disconnection is intentional. The state has special terms for people who use resources without permission and for land not conforming to the dominant model. They are wild settlers, poachers or squatters, they are traditional or backward, and their lands are wastelands. Landscapes are cleaned up of their complexity, and of their natural and social diversity. Hedgerows and ponds are removed, but so are troublesome tribes and the poorest groups. In these landscapes, both real and meta-phorical commons exist. Most of the 700,000 villages of India have, or had, commons – officially designated by name, but vital sources of food, fuel, fodder and medicines for many local people. In northern Europe, open-field or common farming sustained communities for millennia; in southern Europe, huge tracts of uplands are still commonly grazed. In

England and Wales, there are still more than 8000 commons, covering 0.5 million *hectares*, each embodying permanence in the landscape and continuity over generations. Most are archaic reminders of another age in an increasingly industrialized landscape.

Recent thinking and policy has separated food and farming from nature, and then accelerated the disconnectedness. At the same time as real commons have been appropriated, by *enclosures* or prairie expansion, the metaphorical food commons have also been stolen away. Food now largely comes from dysfunctional production systems that harm environments, economies and societies; and yet we seem not to know, or even to care overmuch. The environmental and health costs of losing touch are enormous. The consequences of food systems producing anonymous and homogeneous food are obesity and diet-related diseases for about one tenth of the world's people, and persistent poverty and hunger for another seventh.

So, does sustainability thinking and practice have anything to offer? Can it help to reverse the loss of trust so commonly felt about food systems, and prevent the disappearance of landscapes of importance and beauty? Can it help to put nature and culture back into farming? Can it help to produce safe and abundant food? These are some of the questions addressed in this book, which I believe concern agriculture's most significant revolution. Several themes will reoccur. One is that accumulated and traditional knowledge of landscapes and nature is intimate, insightful and grounded in specific circumstances. Communities sharing such knowledge and working together are likely to engage in sustainable practices that build local renewable assets. Yet, industrialized agriculture, also called *modernist* in this book because it is single coded, inflexible and monocultural, has destroyed much place-located knowledge. In treating food simply as a commodity, it threatens to extinguish associated communities and cultures altogether by conceiving of nature as existing separately from humans. Natural landscapes and sustainable food production systems will only be recreated if we can create new knowledge and understanding, and develop better connections between people and nature.

## The World Food Problem

But why should this idea of putting nature and culture back into agriculture matter? Surely we already know how to increase food production? In developing countries, there have been startling increases in food production since the beginning of the 1960s, a short way into the most recent agricultural revolution in industrialized countries, and just prior to the Green Revolution in developing countries. Since then, total world food production grew by 145 per cent. In Africa, it is up by 140 per cent, in Latin America by almost 200 per cent, and in Asia by a remarkable 280 per cent. The greatest increases have been in China – an extraordinary fivefold increase, mostly occurring in the 1980s and 1990s. In the industrialized regions, production started from a higher base. Yet in the US, it still doubled over 40 years, and in western Europe grew by 68 per cent.[1]

Over the same period, world population has grown from 3 to 6 billion.[2] Again, per capita agricultural production has outpaced population growth. For each person today, there is an extra 25 per cent of food compared with people in 1961. These aggregate figures, though, hide important differences between regions. In Asia and Latin America, per capita food production has stayed ahead, increasing by 76 and 28 per cent respectively. Africa, however, has fared badly, with food production per person 10 per cent less today than in 1961. China, again, performs best, with a trebling of food production per person over the same period. Industrialized countries as a whole show similar patterns: roughly a 40 per cent increase in food production per person.

Yet, these advances in aggregate productivity have only brought limited reductions in incidence of hunger. At the turn of the 21st century, there were nearly 800 million people who were hungry and who lacked adequate access to food, an astonishing 18 per cent of all people in developing countries. One third are in East and South-East Asia, another third in South Asia, a quarter in sub-Saharan Africa, and one twentieth each in Latin America and the Caribbean, and in North Africa and the Near East. Nonetheless, there has been progress to celebrate. Incidences of under-nourishment stood at 960 million in 1970, comprising one third of people in developing countries at the time. Since then, average per capita consumption of food has increased by 17 per cent to 2760 kilocalories per day – good as an average, but still hiding a great many people surviving on less (33 countries, mostly in sub-Saharan Africa, still have per capita food consumption under 2200 kilocaleries per day). The challenge remains huge.[3]

There is also significant food poverty in industrialized countries. In the US, the largest producer and exporter of food in the world, 11 million people are food insecure and hungry, and a further 23 million are hovering close to the edge of hunger – their food supply is uncertain but they are not permanently hungry. Of these, 4 million children are hungry, and another 10 million are hungry for at least one month each year. A further sign that something is wrong is that one in seven people in industrialized countries is now clinically obese, and that five of the ten leading causes of death are diet related – coronary heart disease, some cancers, stroke, diabetes mellitus, and arteriosclerosis. Alarmingly, the obese are increasingly outnumbering the thin in some developing countries, particularly in Brazil, Chile, Colombia, Costa Rica, Cuba, Mexico, Peru and Tunisia.[4]

So, despite great progress, things will probably get worse for many people before they get better. As total population continues to increase, until at least the latter part of the 21st century, so the absolute demand for food will also increase. Increasing incomes will mean that people will have more purchasing power, and this will increase demand for food. But as our diets change, so demand for the types of food will also shift radically. In particular, increasing urbanization means people are more likely to adopt new diets, particularly consuming more meat and fewer traditional cereals and other foods – what Barry Popkin calls the *nutrition transition*.[5]

One of the most important changes in the world food system will come from an increase in the consumption of livestock products. Meat demand is expected to double by 2020, and this will change farming systems.[6] Livestock are important in mixed production systems, using foods and by-products that would not have been consumed by humans. But, increasingly, farmers are finding it easier to raise animals intensively and feed them with cheap cereals. Yet, this is very inefficient: it takes 7 kilogrammes of cereal to produce 1 kilogramme of feedlot beef, 4 kilogrammes to produce one of pork, and 2 kilogrammes to produce one of poultry. This is clearly inefficient, particularly as alternative and effective grass-feeding rearing regimes do exist.[7]

These dietary changes will help to drive a total and per capita increase in demand for cereals. The bad news is that food-consumption disparities between people in industrialized and developing countries are expected to persist. Currently, annual food demand in industrialized countries is 550 kilogrammes of cereal and 78 kilogrammes of meat per person. By contrast, in developing countries, it is only 260 kilogrammes of cereal and 30 kilogrammes of meat per person. These gaps in consumption ought to be deeply worrying to us all.

## Commons and Connections

For most of our history, the daily lives of humans have been played out close to the land. Since our divergence from apes, humans have been hunter-gatherers for 350,000 generations, then mostly agriculturalists for 600, industrialized in some parts of the world for 8 to 10, and lately dependent on industrialized agriculture for just 2 generations.[8] We still have close connections to nature. Yet, many of us in industrialized countries do not have the time to realize it. In developing countries, many are still closely connected, yet are tragically locked into poverty and hunger. A connectedness to place is no kind of desirable life if it brings only a single meal a day, or children unable to attend school for lack of food and books, or options for wage earning that are degrading and soul destroying.

For as long as people have managed natural resources, we have engaged in forms of collective action. Farming households have collaborated on water management, labour sharing and marketing; pastoralists have co-managed grasslands; fishing families and their communities have jointly managed aquatic resources. Such collaboration has been institutionalized in many local associations, through clan or kin groups, water users' groups, grazing management societies, women's self-help groups, youth clubs, farmer experimentation groups, church groups, tree associations, and labour-exchange societies.

Through such groups, constructive resource management rules and norms have been embedded in many cultures – from collective water management in Egypt, Mesopotamia and Indonesia to herders of the Andes and dryland Africa; from water harvesting in Roman North Africa and south-west North America to shifting agriculture systems of the forests of Asia and Africa; and from common fields of Europe to the *iriaichi* in Japan. It has been rare, prior to the last decade or so, for the importance of these local institutions to be recognized in agricultural and rural development. In both developing and industrialized countries, policy and practice have tended to be preoccupied with changing the behaviour of individuals rather than of groups or communities – or, indeed, with changing property regimes – because traditional commons management is seen as destructive. At the same time, modern agriculture has had an increasingly destructive effect on both the environment and rural communities.[9]

A search through the writings of farmers and commentators, from ancient to contemporary times, soon reveals a very strong sense of connectedness between people and the land. The Roman writer Marcus Cato, on the opening page of his book *Di Agri Cultura*, written 2200 years ago, celebrated the high regard in which farmers were held:

> . . . *when our ancestors . . . would praise a worthy man their praise took this form: 'good husbandman', 'good farmer'; one so praised was thought to have received the greatest commendation.*

He also said: '*a good piece of land will please you more at each visit*'. It is revealing that Roman agricultural writers such as Cato, Varro and Columella spoke of agriculture as two things: *agri* and *cultura* (the fields and the culture). It is only very recently that we have filleted out the culture and replaced it with commodity.[10]

It is in China, though, that there is the greatest and most continuous record of agriculture's fundamental ties to communities and culture. Li Wenhua dates the earliest records of integrated crop, tree, livestock and fish farming to the Shang-West Zhou Dynasties of 1600–800 BC. Later, Mensius said in 400 BC:

> *If a family owns a certain piece of land with mulberry trees around it, a house for breeding silkworms, domesticated animals raised in its yard for meat, and crop fields cultivated and managed properly for cereals, it will be prosperous and will not suffer starvation.*

In one of the earliest recognitions of the need for the sustainable use of natural resources, he also said:

> *If the forests are timely felled, then an abundant supply of timber and firewood is ensured; if the fishing net with relatively big holes is timely cast into the pond, then there will be no shortage of fish and turtle for use.*

Still later, other treatises such as the collectively written *Li Shi Chun Qiu* (239 BC) and the *Qi Min Yao Shu* by Jia Sixia (AD 600) celebrated the fundamental value of agriculture to communities and economies, and documented the best approaches for sustaining food production without damage to the environment. These included rotation methods and green manures for soil fertility, the rules and norms for collective management of resources, the raising of fish in rice fields, and the use of manures. As Li Wenhua says: *'these present a picture of a prosperous, diversified rural economy and a vivid sketch of pastoral peace'.*[11]

But it was to be Cartesian reductionism and the enlightenment that changed things many centuries later, largely casting aside the assumed folklore and superstitions of age-old thinking. A revolution in science occurred during the late 16th and 17th centuries, largely due to the observations, theories and experiments of Francis Bacon, Galileo Galilei, René Descartes and Isaac Newton, which brought forth mechanistic reductionism, experimental inquiry and positivist science.[12] These methods brought great progress, and continue to be enormously important. But an unfortunate side effect has been a sadly enduring split, in at least some of our minds, between humans and the rest of nature.

As I discuss later, wilderness writers, landscape painters, ecologists and farmers of the 19th and 20th centuries sought to reverse, or at least temper, the dominance of the new thinking. But it has been a Sisyphean struggle, until perhaps recently when the mountain-top has at least become more visible. It is, though, in the indigenous groups of the world that we find remnants of nature–people connectivity. One of the most comprehensive collections on the diversity of human cultures and their connectedness with nature and the land is Darrell Posey's 700-page volume, *Cultural and Spiritual Values of Biodiversity*. Containing contributions from nearly 300 authors from across the world, these highlight '*the central importance of cultural and spiritual values in an appreciation and preservation of all life*'. These voices of the earth demonstrate the widespread intimate connectivity that people have with nature, and their mutual respect and understanding.[13]

In Australia, Henrietta Fourmile of the Polidingi Tribe says:

*Not only is it the land and soil that forms our connections with the earth but also our entire life cycle touches most of our surroundings. The fact that our people hunt and gather these particular species on the land means emphasis is placed on maintaining their presence in the future. . . What is sometimes called 'wildlife' in Australia isn't wild; rather, it's something that we have always maintained and will continue gathering.*

Pera of the Bakalaharil tribe in Botswana points to their attitudes in using and sustaining wild resources:

*Some of our food is from the wild, like fruits and some of our meat. . . We are happy to conserve, but some conservationists come and say that preservation means that we cannot use the animals at all. To us, preservation means to use, but with love, so that you can use again tomorrow and the following year.*[14]

Johan Mathis Turi of the Saami reflects on the mutual shaping in the Norwegian Arctic:

*The reindeer is the centre of nature as a whole and I feel I hunt whatever nature gives. Our lives have remained around the reindeer and this is how we have managed the new times so well. It is difficult for me to pick out specific details or particular incidences as explanations for what has happened because my daily life, my nature, is so comprehensive. It includes everything. We say 'lotwantua', which means everything is included.*

A similar perspective is put by Gamaillie Kilukishah, an Inuit from northern Canada who, in translation by Meeka Mike, says:

*You must be in constant contact with the land and the animals and the plants. . . When Gamaillie was growing up, he was taught to respect animals in such a way as to survive from them. At the same time, he was taught to treat them as kindly as you would another fellow person.*

This Inuit perspective is common across the Arctic. Fikret Berkes documents the careful management by the Cree of the Canadian eastern sub-Arctic populations of beaver, caribou and fish. None of the species used by the Cree has become locally extinct since the glaciers departed the region some 4000–5000 years ago. Berkes says: '*hunters are experts on the natural history of a number of species, and on food chain and habitat relationships*'. The management of beaver is particularly clever. Cree communities appoint stewards, or beaver bosses, who oversee the codes and rules for hunting and are also a chief source of knowledge about past hunting patterns and current beaver abundance. The trick is to manage in balance. If there are too many beavers, and the willow and aspen decrease until a threshold is passed, beaver numbers crash, and the whole system takes many years to recover. Cree management involves hunting once every four years to prevent such an ecosystem flip. Berkes indicates the subtle way the Cree see this balance: '*these adjustments are articulated in terms of the principle that it is the animals (and not the hunter) who are in control of the hunt*'. Thus, there is reciprocity between animal and hunter, and these connections echo similar rules for social relations. For the Cree, there is no fundamental difference between people and animals.[15]

Some believe that the ruin of common resources is inevitable – an unavoidable tragedy, as Garrett Hardin put it more than 30 years ago.[16] Each person feels compelled to put another cow on the common, because each derives all the benefit from the additional animal; but the costs are distributed amongst all of the other common users. In the contemporary context, each polluter continues to add greenhouse gases to the atmosphere, while reaping the immediate benefit of not having to pay the cost of abating the pollution, or of adopting clean practices. The costs, though, are spread amongst us all – including future generations who will have to pay for climate change. Other theorists have been equally pessimistic. Mansur Olson was convinced that unless there is coercion or individual inducements, then '*rational, self-interested individuals will not act to achieve their common or group interest*'.[17] This indicates a problem with free riders – individuals who take the benefit, but do not invest anything in return. The temptation, some would say, is always to free ride. The logic has been so compelling that the state has stepped in, developing policies directly or indirectly to privatize common property systems. Although this has been going on for centuries, it has accelerated during the late 20th century' experiment with modernism. Yet, productive commons persist in many parts of the world.

In some places, the loss of local institutions has led to further natural resource degradation. In India, management systems for *common property resources* have been undermined, a critical factor in the increased over-exploitation, poor upkeep, and physical degradation observed over the past half century. As local institutions have disappeared, so the state has felt obliged to take responsibility for natural resources, largely because of a mistaken assumption that resources are mismanaged by local people. This solution is rarely beneficial for environments or for poor people. A key question, therefore, centres on how can we avoid this double tragedy of the commons, in which both nature and community are damaged. It is in precisely this area that there have been so many heroic transformations, and why there is increasing hope now for a new future for agricultural and food systems.[18]

## On Shaping and Being Shaped

Some may feel there is little value in connecting us to the land and nature. Is it not just something for indigenous people or remote tribes? What possible meaning or value can come from an abstract idea such as connectedness to nature? Firstly, even in our modern times, we as predominantly urban-based societies never seem to get enough of nature. People in cities and towns are wistful about lost rural idylls. They visit the countryside on Sunday afternoons, or for occasional weekends, but on returning home, often feel that they should have stayed. Membership of environmental organizations in industrialized countries has never been higher and is growing. In many developing countries, city people do not just go to rural areas for the experience – they return to their home farms. If you ask urban dwellers in cities from Nairobi to Dakar: 'where do you live?', they likely as not will give the name of their rural village or settlement rather than the city. Their family still farms; they earn in the city to invest in the farm and its community. Here the connectedness is tangible.

Yet, an intimate connection to nature is both a basic right and a basic need. When it is taken away, we deny it was ever important, or simply substitute occasional visits and personal experiences. But it is still there, and it is valuable. Is it any wonder to discover that the gentle opportunities afforded by urban community gardens have brought meaning and peace to many people with mental health problems? For all of our time, we have shaped nature, and it has shaped us, and we are an emergent property of this relationship. We cannot suddenly act as if we are separate. If we do so, we simply recreate the wasteland inside of ourselves.

The world we see, through our window, or from space above, is shaped by us. From a distance, it is, of course, larger than any obvious shaping. But this is not our scale. Our scale is more local, though the effects are often greater. What we see around us has been shaped by us. Agricultural landscapes are obviously created, whether rice terraces upon an Asian hillside, or prairie farms in the North American plains, or rolling European patchwork fields. But even most 'natural' or 'wild' landscapes are also creations of this interaction. Few forests are truly pristine wildernesses. Most arise from some human shaping, even the Amazon rainforests and the northern tundra. Strangely, most contemporary debates on human–nature interactions focus on how nature has been shaped by us, without fully accepting the second part of the equation: that we, too, must be shaped by this connection, by nature itself.

We are also shaped by our systems of food production, as they, in turn, shape nature, and rely upon its resources for success. We are affected by what we know about these systems – whether we approve or disapprove, whether the food system is local or distant. We are, of course, fundamentally shaped by the food itself. Without food, we are clearly nothing. It is not a lifestyle add-on or a fashion statement. The choices we make about food affect both us, intrinsically, and nature, extrinsically. We make one set of choices, and we end up with a diet-related disease and a damaged environment. We make another set, and we eat healthily, and sustain nature through sustainable systems of food production. In truth, it is not such a simple dichotomy as this. But once we accept the idea of the fundamental nature of this connection, then we start to see options for personal, collective and global recovery.

The connection is philosophical, spiritual and physical. We are buying a system of production when we purchase its food. In effect, we eat the view and consume the landscape. Clearly, the more we consume of one thing, the more it is likely to be produced. But if the system of production has negative side effects, and cares not about the resources upon which it relies, then we have taken a path leading, ultimately, to disaster. On the other hand, if our choices mean more food comes from systems of agricultural production that increase the stock of nature, that improve the environment while at the same time producing the food, then this is a different path – a path towards sustainability. We must now shape this new path. We will, by walking it, also change ourselves. We will adapt and evolve, and new connections will be established.

Nature is amended and reshaped through our connections – both for the bad and for the good. But I am worried, too, as the worst kind of reshaping occurs when nature is destroyed, or ignored, and then recreated in a 'themed' context. Do not worry about the losses, we might be saying, we can make it better than the original. When nature is themed, the outcome is grim: plants and trees are made from plastic, sand is laid down by the millions of tonnes to create new beaches, and rocks are sprayed with cement to look more 'natural'.[19] But this should not diminish the value of nature as an escape, ultimately a mystery, and an 'otherness' from life in the city. It is an imagined world, as well as a real world full of great meaning and significance.

## This Disconnected Dualism

Is nature part of us, and we a part of a grander scheme? Or are we, as humans, somehow separate? These are questions that have exercised philosophers, scientists and theologists through the ages, and particularly since the Enlightenment, when Newton's mechanics and Descartes' 'nature as machine' helped to set out a new way of thinking for Europeans. The result has been the gradual erosion of connections to nature and the emergence, in many people's minds, of two separate entities – people and nature.

During recent years, with growing concerns for sustainability, the environment, and biodiversity, many different typologies have been developed to categorize shades of deep- to shallow-green thinking. Arne Naess sees shallow ecology, for example, as an approach centred on efficiency of resource use, whereas deep ecology transcends conservation in favour of biocentric values. Other typologies include Donald Worster's imperial and Arcadian ecology, and the resource and holistic schools of conservation. For some, there is an even more fundamental schism: whether nature exists independently of us, or whether it is characterized as post-modern or as part of a post-modern condition. Nature to scientific ecologists exists. To post-modernists, though, it is all a cultural construction. The truth is, surely, that nature does exist, but that we socially construct its meaning to us. Such meanings and values change over time, and between different groups of people.[20]

There are many dangers in the persistent dualism that separates humans from nature. It appears to suggest that we can be objective and independent observers, rather than part of the system and inevitably bound up in it. Everything we know about the world, we know because we interact with it, or it with us. Thus, if each of our views is unique, we should listen to the accounts of others and observe carefully their actions. Another problem is that nature is seen as having boundaries – the edges of parks or protected areas. At the landscape level, this creates difficulties because the whole is always more important than each part, and diversity is an important outcome.[21]

This leads inevitably to the idea of enclaves – social enclaves such as reservations, barrios or Chinatowns, and natural enclaves such as national parks, wildernesses, sites of special scientific interest, protected areas or zoos. Enclave thinking leads us away from accepting the connectivity of nature and people. It appears to suggest that biodiversity and conservation can be in one place, and productive agricultural activities in another.[22] So, is it acceptable to cause damage in most social and natural landscapes, provided you leave a few tasty morsels at the edges? Surely not. These enclaves will always be under threat at the borders, or simply be too small to be ecologically or socially viable. They also act as a sop to those with a conscience – we can justify the wider destruction if we fashion a small space in which natural history can persist.

By continuing to separate humans and nature, the dualism also appears to suggest that we can invent simple technologies that intervene to reverse the damage caused by this very dualism. The greater vision, and the more difficult to define, involves looking at the whole and seeking ways to redesign it. The Cartesian 'either/or' between humans and nature remains a strange concept to many human cultures. It is only modernist thinking that has separated humans from nature in the first place, putting us up as distant controllers. Most peoples do not externalize nature in this way. From the Ashéninha of Peru to the forest dwellers of former Zaire, people see themselves as just one part of a larger whole. Their relationships with nature are dialectical and holistic, based on 'both/with' rather than 'either/ or'.[23]

For the Arakmbut of the Peruvian rainforest, Andrew Gray says: *'no species is isolated; each is part of a living collectivity binding human, animal and spirit'*. Mythologies and rituals express and embed these inter-relationships, both at the practical level, such as through the number of animals a hunter may kill and how the meat should be shared, and at the spiritual level, in which *'the distinction between animal, human and spirit becomes blurred'*. One of the best known of these visible and invisible connections is the Australian Aboriginal peoples' Dreamtimes. Aboriginal people have inhabited Australia for 30,000 years or more, during which time some 250 different language groups developed intimate relations with their own landscapes. David Bennett says:

> *Aboriginal peoples hold that there is a direct connection between themselves and their ancestral beings, and because they hold that their country and their ancestral beings are inseparable, they hold that there is a direct connection between themselves and their country.*[24]

These connections are woven into the Dreamtime, or the Dreaming, which in turn shapes the norms, values and ideals of people within the landscape.

Each Aboriginal group has its own stories about the creation of their land by their ancestors, and these stories connect people with today's land. Such land is non-transferable. It is not a commodity; therefore, it cannot be traded. Events took place here, and people invested their lives and built enduring connections – so no one owns it; or, rather, everyone does. As Bennett also says: *'those who use the land have a collective responsibility to protect, sustainably manage and maintain their "country"'*. How sad that those who came later showed so little of this responsibility and little collective desire to protect what was already present.

# Wilderness Ideas

The idea of the wilderness struck a chord during the mid 19th century, with the influential writers Henry David Thoreau and John Muir setting out a new philosophy for our relations with nature. This grew out of a recognition of the value of wildlands for people's well-being. Without them, we are nothing; with them, we have life. Thoreau famously said in 1851: *'in wildness is the preservation of the world'*. Muir, in turn, indicated that: *'wildness is a necessity; and mountain parks and reservations are useful not only as fountains of timber and irrigating rivers, but as fountains of life'*. But, as Roderick Nash, Max Oelschlaeger, Simon Schama and many other commentators have pointed out, these concerns for wilderness represented much more than a defence of unencroached lands.[25] They involved the construction of a deeper idea – an imagination of something that never really existed, but which proved to be hugely successful in reawakening, in North American and European consciences, the fundamental value of nature.

Debates have since raged over whether 'discovered' landscapes were 'virgin' lands or 'widowed' ones, left behind after the death of indigenous peoples. Did wildernesses exist, or did we create them? Donald Worster, environmental historian, points out for North America that *'neither adjective will quite do, for the continent was far too big and diverse to be so simply gendered and personalized'*.[26] In other words, just because they constructed this idea does not mean to say it was an error. Nonetheless, they were wrong to imply that the wildernesses in, say, Yosemite were untouched by the human hand; these landscapes and habitats were deliberately constructed by Ahwahneechee and other Native Americans and their management practices in order to enhance valued fauna and flora.

Henry David Thoreau developed his idea of people and their cultures as being intricately embedded in nature as a fundamental critique of mechanical ideas that had separated nature from its observers. His was an organic view of the connections between people and nature.[27] In his *Natural History*, Thoreau celebrates learning by '*direct intercourse and sympathy*', and advocates a scientific wisdom that arises from local knowledge accumulated from experience, combined with the science of induction and deduction. However, he still invokes the core idea of wilderness as untouched by humans, even though his home state of Massachusetts had been colonized just two centuries earlier and had a long history of 'taming' both nature and local Native Americans.

Nature is something to which we can escape as individuals. Thoreau celebrates the rhythms of walking and careful observation. He:

> . . .*looked with awe at the ground. . . Here was no man's gardens, but the unhandselled globe. It was not lawn, nor pasture, not mead, nor woodland, not lea, nor arable, not wasteland. It was the fresh and natural surface of the planet Earth, as it was made forever and ever.*

The important thing to note here is that the elegiac narrative of connections and intrinsic value had a huge influence on readers; and perhaps it is a small price to pay that Thoreau focused on the 'unhandselled globe' and the 'fresh and natural' to the exclusion of other constructed natures. For these woods were, of course, shaped in some way by previous peoples – they are an outcome of both people and nature, not a remnant of primary wilderness until he happened along.[28]

The question 'is a landscape wild, or is it managed' is perhaps the wrong one to ask, as it encourages unnecessary and lengthy argument. What is more important is the notion of human intervention in a nature of which we are part. Sometimes such intervention means doing nothing at all – leaving a whole landscape in a 'wild' state – or perhaps it means just protecting the last remaining tree in an urban neighbourhood or a hedgerow on a field boundary. Preferably, intervention should mean sensitive management, with a light touch on the landscape. Or it may mean heavy reshaping of the land, for the good or the bad.

So, it does not matter whether untouched and pristine wildernesses actually exist. Nature exists without us; with us it is shaped and reshaped. Most of what exists today does so because it has been influenced explicitly or implicitly by the hands of humans, mainly because our reach has spread as our numbers have grown, and because our consumption patterns have compounded the effect. But there are still places that seem truly wild, and these exist at very different scales and touch us in different ways. Some are on a continental scale, such as the Antarctic. Others are entirely local: a woodland amidst farmed fields, a salt marsh along an estuary, a mysterious urban garden – all touched with private and special meanings.

This suggests that wild nature and wilderness can exist on a personal scale. If we find a moment's peace on an hour's walk across a meadow by the river, does it matter that this is a shaped nature, and not a wild one? Wilderness is an idea, and it is a deep and appealing one. Some shaping of landscape can be so subtle that we hardly notice. Nigel Cooper asks how natural is a nature reserve, and identifies a range of places where conceptions of nature are located in the British landscape, including biodiversity reserves, wilderness areas, historic countryside parks, and what he calls 'companion places'. In our almost entirely farmed landscape, where nature is as much a product of agriculture as it is an input, the efforts to recognize and conserve biodiversity and wilderness are varied. All of these are as much treasured by the people who make or experience them as those who gaze upon the wildest forests, savannahs or mountains.[29]

In all of these situations, we are a part, connected; we affect nature and land, and are affected by it. This is a different position to one which suggests that wilderness is untouched, pristine, and so somehow better because it is separated from humans – who, irony of ironies, promptly want to go there in large numbers precisely because it appears separate. But an historical understanding of what has happened to produce the landscape or nature we see before us matters enormously when we use an idea to form a vision that clashes with the truth. One idea may be that a place is wild, and so local people should be removed from it. Another idea is that a place is ripe for development, and so a group of people should be dispossessed. The term wilderness has come to mean many things, usually implying an absence of people and the presence of wild animals; but it also contains something to do with the feelings and emotions that are provoked in people. Roderick Nash takes a particularly Eurocentric perspective in saying *'any place in which a person feels stripped of guidance, lost and perplexed may be called a wilderness'*, though this definition may also be true of some harsh urban landscapes.[30] The important thing is not defining what it really is, but what we think it is, and then telling stories about it.

## Stories and Memories of the Wild

The landscape is full of stories and meanings that we have made of stones or trees, of plants and rivers, memories that we have woven together with beginnings and ends. This creativity gives extra life to nature and how we react to it, and how we are shaped by it. But how good are we at still telling these stories? Ben Okri, in his *Joys of Story Telling*, says of Africa:

*Everything is a story, everything a repository of stories, spiders, the wind, a leaf, a tree, the moon, silence, a glance, a mysterious old man, an owl at midnight, a sign, a white stone on a branch, a single yellow bird of omen, an inexplicable death, unprompted laughter, an egg by the river, are all impregnated with stories. In Africa, things are stories, they store stories, and they yield stories at the right moment of dreaming, when we are open to the secret side of objects and moods.*[31]

I have my own African stories, also having been born in Nigeria, and then spending my formative childhood years there. Thinking about that time, I realize that many of my most vivid early memories are of encounters with animals. Perhaps it is that way with all children, or perhaps it was the place. I recall meeting a snake in the bathroom, chancing on a lion while walking in the bush, being chased by a scorpion in a long-empty swimming pool. I remember huge rats downed with a shot gun, and tail to jaw as long as I was tall, and a ferocious serval cat prowling on the roof until it, too, was shot. There were songbirds in the aviary, large dogs, monkeys as pets, itinerant donkeys, and great silent fruit bats at dusk. Some of these memories could be no more than childhood constructions, though flickering reels of super-8 film still testify to many truths.

Whether Africa has more stories than another place, or even too many, as Okri hints, matters less than the fact that industrialized landscapes have lost many of their stories. We no longer see the deep significance; we no longer know the old ways. Many of these are dark and well worth forgetting. But the stories we have written on the industrialized landscape in recent decades have been bad, perhaps much worse. There is meaning in the landscape, and as Oliver Rackham has put it: *'I am especially concerned with the loss of memory. The landscape is a record of our roots and the growth of civilisation'.*[32]

Many writers have suggested that we are in between ages, on the point of discovery or rediscovery. We have forgotten so much about human linkages with the rest of nature, and about our fundamental dependencies. David Suzuki says:

*We feel ourselves to have escaped the limits of nature. . . Food is often highly processed and comes in packages, revealing little of its origins in the soil or tell-tale signs of blemishes, blood, feathers or scales. We forget the source of our water and energy, the destination of our garbage or our sewage. We forget.*

Here is something vital. When we forget these truths, we come easily to believe another story – that we have the Earth under our control. Suzuki says: *'we must find a new story'.*[33] This has been Thomas Berry's mission, too. He says: *'It's all a question of story. We are in trouble just now because we do not have a good story. We are in between stories. The old story, the account of how we fit into it, is no longer effective. Yet we have not learned the new story.*[34] In East Anglia, home to one of the two giants of working horses, the Suffolk Punch, horsemen looked upon the landscape and saw it full of wild plants with vital uses. Today, the horses have gone, replaced by tractors and combines, and the useful plants are merely weeds. We have forgotten. Perhaps this is progress? Or perhaps we have to find new ways of valuing, using and constructing the nature around us?

It is sad that so much knowledge of nature, its uses and significance, has slipped away; such stories take time to build. They arise from the experiences of the many, from the insights of a few, and from the sharing of such significance. When we no longer find the need or desire to tell stories about nature, then the thread is broken. That, of course, plays into the hands of those who would cut down the tree, or pollute the water, or allow the soils to slide into the river. But where there are collective connections, through farmers working together, or consumers linked directly to a farm, or walkers strolling together across a landscape, then it is possible to create new stories. Perhaps it is possible even to rediscover some of those stories assumed now to be lost. The problem is that there, strangely, still persists amongst many of us a dislocation between traditional knowledge of land and nature and what we might term modern scientific knowledge. We commonly hold apparently conflicting knowledge side by side without feeling particularly harmed – often, in fact, within the same scientific discipline.

A decade ago, on a training course in Kenya for government officers, I asked participants to list examples of their traditional knowledge of nature. Our intention was to encourage highly trained professionals to reflect on the value of the knowledge and insights of local people – not to say that it was better knowledge; just that it was worth listening to and incorporating with other, more scientific, sources. A remote community cannot know the detailed mechanisms by which legumes interact with rhizobia in the soil to fix atmospheric nitrogen; nor will they know the properties of a chemical that pollutes a well. What they know will have been built up from accumulated individual and collective experiences, fixed in time through story-telling. In this one session, baking beneath the hard equatorial sun, we listed more than 40 well-accepted idioms, ideas and stories. Many were to do with trees. In some places, the bark of *Acacia* is used to treat malaria, and its ash to cure milk and give it good taste. Ash from other trees is sprinkled on crops to control various pests and weeds. One tree, *Croton*, is not permitted to grow near houses because of the belief that someone will die if roots enter the house. Elsewhere, the wood from certain trees is never used for beds, as it is believed to make women infertile. The *Erythrina* tree is accorded magical properties in curing mumps in children. It is true that some of these ideas are just superstitions – stories without a sound empirical base. But drawing the line between what may only be an unfounded superstition and something with more than a degree of truth is not easy.

In Britain, most folklore about plants and animals has its roots in Celtic, Roman, Anglo-Saxon and Norse traditions that date from 1000 to 2000 years ago. Some trees are associated with magic, such as the oak, and others with repelling witches, such as the holly (or iron tree), which is also a protector against lightning and fires. Other important trees include elder (a sacred tree in Celtic religion), ash (well known for curing illnesses), birch (for protection against witches), hawthorn (for good milk yields and lightning protection), and yew (with its associations with death as a graveyard plant).

In his *Flora Britannica*, Richard Mabey suggests that at least 500 churchyards in England and Wales, out of 12,000 surviving churches, contain yew trees that are at least as old as the church itself.[35] Such ancient trees, living for up to 1000 years, are associated with the accumulation of many memories and customs. The oak, of course, has special signif-icance: the shanty 'heart of oak are our ships, heart of oak are our men' has been part of national folklore for over two centuries. Oak leaves garland and disguise carvings of the pagan 'green man', still seen in many churches. Some uses of plants are tied to collective and family customs, particularly gathering bilberry, also known as whortleberry and whinberry, with families from Devon, Somerset, Shropshire, Surrey, the Isle of Man and the Pennines travelling up to the moors in August to gather berries for home use. People engaged in these activities take great care. One west country woman says: *'we gathered it carefully, not haphazardly, remembering there was a tomorrow'.*

Wild plants may no longer have livelihood relevance; without them, most Britons would not suffer hunger, or lack for medicines. But they still retain an encouragingly deep cultural significance. We may buy plastic-wrapped fruit and microwave meals, where food is not much more than a commodity; but many wild plants still have a wider significance. As Mabey puts it:

> *Plants have had symbolic significance as well as utilitarian meanings since the beginnings of civilization. They have been tokens of birth, death, harvest, and celebration, and omens of good (and bad) luck. They are powerful emblems of place and identity, too, not just of nations, but of villages, neighbourhoods, even personal retreats.*

Holly and mistletoe carry magical meanings, and are associated with both pagan and Christian festivals. May blossom brings bad luck when brought into the house; daisies are fashioned into chains by children; algae and pine cones are used for weather forecasting; weld and woad for yellow and blue dyes; samphire for glass-making; silverweed for aching feet; junipers and sloes for flavouring gin; hellebores for treating worms, and nettles for arthritis; and holly wood is valued for its power over horses. Look hard, and it is possible to find traditions associated with most plants and animals – some strange and mythical, others with more obvious empirical truths.[36]

The knowledge we have about plants and animals is extraordinary. They are a connection between us and place, between memory and identity, between myth and meaning. Not all are traditional or old. Recent years have, for example, seen the widespread use of young crack willow to make living seats and cribs, and red poppies are worn to remember war casualties, originating from World War I. The important thing is that plants and animals play roles in culture beyond those of obvious economic purpose. But when the plant is lost forever, the tree is cut down or the weed removed, then the culture associated with it goes, too. Equally, when the cultural knowledge disappears, or is replaced with something else, then another reason for preserving biological diversity is lost. It is sad that so many rural customs and festivals no longer carry any significance in our modern world. At one time, they were a central part of community life.[37]

But such diversity of knowledge and meaning can only arise when the landscape is itself diverse. A *monoscape* of highly controlled and large-scale farming has no room for wild foods or their cultural significance. It neither wants them nor needs them. So what is lost when they go is not just a weed or two. It is something of a culture – a connection between people and land lost forever, save for a few lingering memories in dusty books.

## Language and Memory on the Frontier

Many stories about nature and our Earth are embedded in local languages. Language and land are part of people's identities, and both are under threat. There are 5000–7000 oral languages spoken today, only about a half of which have more than 10,000 speakers each.[38] The rest, about 3400 languages, are spoken by only 8 million people, about one tenth of 1 per cent of the world's population. The top ten spoken languages now comprise about half of the world's population. A great deal of linguistic diversity is thus maintained by a large number of small and dwindling communities. They, like their local ecologies and cultural traditions, are under threat. Here, there is a vicious circle. As languages come under threat, so do the stories that people tell about their environments. Local knowledge does not easily translate into majority languages, and moreover, as Luisa Maffi states: *'along with the dominant language usually comes a dominant cultural framework which begins to take over'.*[39]

Thus, we increasingly lack the capacity to describe changes to the environment and nature, even if we are able to observe them. Slowly, it all slips away. Gary Nabhan and colleagues describe how the children of the Tohono O'odham (formerly known as the Papago) of the Sonoran Desert in the south-west US are losing both a connection with the desert and with their language and culture. Even though they hear the language spoken at home, they are not exposed to traditional story-telling, and are no longer able to name common plants and animals in O'odham – though they could easily name large animals of the African savanna seen on television. Nabhan called this process of erosion the 'extinction of experience'.[40]

These losses, too, are hastened by land degradation or removal for other purposes. Also in the Sonoran Desert, Felipe Molina found that his own people, the Yaqui or Yoeme, were unable to perform traditional rituals because of the disappearance of many local plants. Land is being settled by non-Yoeme and converted to other uses.[41] Biodiversity slips away, and only the local indigenous people notice. But they are powerless in the global scheme of things. Their intimate spiritual and physical connection with nature is under threat; yet, we on the outside may never notice it disappear.

> *Yaquis have always believed that a close communication exists among all the inhabitants of the Sonoran Desert world in which they live: plants, animals, birds, fishes, even rocks and springs. All of these come together as part of one living community which Yaquis call the huya ania, the wilderness world.*[42]

These problems are all connected. Luisa Maffi adds:

> *The Yoeme elders' inability to correctly perform rituals due to environmental degradation thus contributes to precipitating language and knowledge loss and creates a vicious circle that in turn affects the local ecosystem.*[43]

The concept of the frontier suggests to me a place where people test out existing ideas on a new environment. As a result, both change. William Cronon and fellow historians indicate that self-shaping occurs rapidly on the frontier. The different identities of groups arriving from distant places, and those of people already present, clash and blend, merge and stand apart. Of the American frontier, they say: *'Self-shaping was a part of the very earliest frontier encounters and continues as a central challenge of regional life right down to the present.*' People on the Western frontier, as they pushed into what they saw as a 'wilderness' and 'free-land', had *borrowed most of their cultural values. . . from Europe and older settlements back east'.* They reshaped nature and themselves. They also, of course, imposed a new landscape on the old. Through conquest, the original owners were removed and corralled. New stories and mythologies emerged to give greater justification to these acts. One set of ideas about a landscape was replaced by another.[44]

The pioneering frontier historian, Frederick Jackson Turner, though promoting many ideas and views long since shown to be wrong and even downright racist, rightly indicated that the frontier repeated itself.[45] The frontier, where shaping of nature and self-shaping of societies are combined with a destruction of existing relationships and cultures, expands today at a pace beyond the appreciation of the majority. Most shaping does not bring benefits to us all, as the interwoven rug of nature and people is steadily pulled from beneath our feet. I am not concerned here with defining exactly what is a frontier or, indeed, where frontiers exist. Its use as an idea lies in the notion that one set of values about a land comes rapidly to be imposed upon another. In modern times, the frontier is characterized both by the expansion of modern industrialized agriculture, or by the loss of local associations and connectivity to the land. The problem is that those pushing out the frontier see it as progress; those exposed to the invasion see mainly destruction and loss. Of course, this applies, too, to the contemporary expansion of sustainable agriculture. When William Bradford stepped off the *Mayflower*, he saw a *'hideous and desolate wilderness'*.[46] The pioneers at the frontier were not only carving out new lives, but battling it out with the wild country for survival. As Nash put it:

> *Countless diaries, addresses and memorials of the frontier period represented wilderness as an 'enemy' which had to be 'conquered', 'subdued' and 'vanquished' by a 'pioneer army'. The same phraseology persisted into the present century.*[47]

In practice, of course, there is always mingling at the frontier, and what we see is a function of both sides' capacity to shape and reshape. Those coming along to the frontier bring connections to old cultures, but also new ideas about how to make improvements. Recipients at the frontier find new opportunities to trade, interact and learn. Out of these new connections can come new forms of cross-cultural dialogue. In the early north-east US, for example, where the received story is one of misunderstanding and conquest, the British and French learnt Iroquois languages, protocols and metaphors in order to aid trust and trading.[48] But it is also true that, in the end, there are clear winners and losers. As land beyond the frontier is seen as 'free', so it is taken, and this inevitably means conflict and violence. Cronon and colleagues say:

> *Sometimes, it was perpetrated by individuals, and sometimes by the military power of the state. Always, it drew dark lines on a landscape whose newly created borders were defeated with bullets, blades and blood.*[49]

Today, such frontier experiences are played out in the rainforests, swamps, hills and mountains of Latin America, Africa and Asia, and in the landscapes overwhelmed by modern agricultural technologies and narratives. What is gained is one thing – more food. What is lost has been too often invisible. Yet what is equally important are the cognitive frontiers inside of ourselves. We each have a journey to travel if we are to find new ways of protecting our world, while at the same time producing the food we need.

## It Does Matter Who Tells the Story

Who gets to tell the stories matters greatly. Every piece of land or landscape contains as many meanings and constructions as the people who have interacted with it. A modern industrialized landscape, let's call it a monoscape, has few meanings. By contrast, a *diverscape* has many. Thus, a single story of the land is not the only story, though many would have us believe it to be true. When the Europeans first brought their visions to the Pacific and Australasia, they saw the landscape and met the people. But they did not give them great value – that is, beyond curiosity and museum value. They sought to save them, convert them, enslave them. They imposed their stories on the landscape – even though Aboriginal peoples in Australia had walked the land for at least 1500 generations, and had accumulated extraordinary knowledge, understanding and compelling stories over time scales beyond any persisting European culture.[50] As Paul Carter describes, Captain Cook and the 'first arrivers' and narrators saw an empty space that could be settled and civilized.[51] The Australian landscape was awaiting history, and new stories could be created and imposed upon others. They named all that they saw – in four months over 100 bays, capes and isles. Carter says *'for Cook, knowing and naming were identical'*. Once these discovered places had been named for the first time, so they were known. The landscape begins its process of being reshaped. Cook sees, on deep black soils, *'as fine a meadow as ever was seen'*. Such meadows were rather like those of home, and echoed John Muir's observation of 'wild' meadows in Yosemite that were actually created by controlled fires set by Native Americans.

The naming of the new, which was actually old, with the old from elsewhere continued apace for decades, as explorers forced their way into the interior, aiming, as Carter put it, to *'dignify even hints of the habitable with significant names. . . Possession of the country depended. . . to some extent, on civilizing the landscape, bringing it into orderly being'*. The new story is told and written, and the old slips away without notice. At the time, few bothered to find out about the local stories of landscape, of the song lines stretching across both thousands of years and thousands of kilometres. Song lines wrap nature and the landscape inextricably into culture, identity and community. Take one away, and the whole falls apart.

Today, 229 years after Cook's landfall, I am standing with Phil and Suzie Grice on their Western Australian wool and cereal farm. They have an ecologically literate view of the landscape. They had seen what happened through modern farming, and where it had led their family and neighbours. In a brief two centuries, modern farming and land management methods brought substantial economic benefit, but great harm, too, to the environment and land. Phil says: *'For two generations, the previous owner and his father pushed back the frontier, removing nature and replacing it with fields. Now, I'm replanting native vegetation as fast as I can and afford'*. The farm is in Lower Balgarup catchment, 260 kilometres south-west of Perth, set in a landscape of ancient and deeply weathered soils. But in the blink of an eye, it has changed. In the 40 years to 1990, 85 per cent of all the natural vegetation in the catchment was removed, with a profound impact on both hydrology and local biodiversity. Soils and water have become salinized, and farming itself threatened. The cost of expansion of the farming frontier has been destruction of the very resource upon which farmers relied.[52]

Eighteen farmers set up the Lower Balgarup catchment group in 1990, covering an area of some 14,000 hectares. It is one of 400 Landcare groups in Western Australia. One of the first actions of the group was to survey the area of land degradation because no one quite knew the extent of the problem. They were shocked to find more than 600 hectares of land affected by dryland salinity and waterlogging. Since then, Phil and his neighbours have planted 200,000 trees, constructed 100 kilometres of new fencing to protect creeks, and another 70 kilometres of drains and banks, and put down land to perennial grasses. The trees and grasses help to pump groundwater by evapotranspiration, so reducing salinity. But the task for the whole landscape is still massive. There are 19 million hectares of wheat and wool country in Western Australia, and already nearly 2 million hectares have been lost to dryland salinity. By 2010, another 3 million hectares are expected to have been lost and 40 rural towns in the wheat belt will have become vulnerable. This ancient landscape, where the rocks of the Yildirim block underlying the catchment are 2.5 billion years old, needs thorough redesigning. Can these farmers, with their changed ways of thinking, now construct a new story?

Of course, what Cook saw, and later Muir in the Sierra Nevada, was conditioned by what they knew. If you believe in wildernesses, then you will see one and name it so. If you know a meadow as part of a pastoral scene, so you will see one more readily. If you see native vegetation simply taking up space where fields could be, then you remove it. However, it is a mistake to believe that the effect of rewriting the landscape is only a one-way process. As Bernard Smith indicates with regard to the 'discovery' of Pacific peoples during the 18th century, their impact on Europe was perhaps as great as the impact of European culture and diseases on the Pacific.[53] When the Tahitian, Omani, arrived in England in 1774 with Captain Furneaux, according to Smith he *'created a sensation. . . He mingled in fashionable circles with a natural grace and became a lion of London society'*. More importantly, his presence provoked new domestic criticism of Empire and its 'pilfered wealth', and even of the shortcomings of English society. A decade later, the son of the chief of one the Palau Islands accompanied Henry Wilson back to England, again to much public acclaim and self-criticism.

Nonetheless, there persisted a subtle misrepresentation of the story through landscape painting that, according to Smith, sought to *'evoke in new settlers an emotional engagement with the land that they had alienated from its aboriginal occupants'*. The noble Pacific islander, in traditional dress, or engaged in traditional ceremony or dance, or the boat full of arriving heroes sensitively stepping onto the beach, hides the real story. Landings were more often accompanied by guns and violence, and long-term damage to societies and nature. Such systematic disenfranchisement has clearly been more common than sensitive interaction. George Miles similarly draws attention to the lack of voice given to Native Americans by incomers. Even though they had told their stories for centuries, suddenly they were silent, nobly silent to some, but more often – sadly even to the likes of Mark Twain – they were *'silent, sneaking, treacherous looking'*.[54]

Part of the problem was that most Native Americans had a predominantly verbal culture, without alphabets. The Cherokee alphabet, for example, was only constructed by a young Cherokee, Sequoyah, in the early 19th century. It led to the printing of the first Native American newspaper in 1828, which was so successful in telling its story that the authorities of Georgia arrested its editors and confiscated the press six years later. It then reappeared as the *Cherokee Advocate* in 1843, from the Cherokee national capital of Talhequah, lasted until 1854, was closed down again, reappeared again in the mid 1870s, and then endured until 1906, when its 800 to 1000 Cherokee-only readers finally lost their only national language paper. During the 18th and 19th centuries, according to Miles, nearly every Native American community embraced opportunities to write and read their own languages: *'from the Micmacs of Newfoundland to the Sioux of the plains, from the Apaches, Navajos and Yaquis of south-west and the Luiseños of California to the Aleuts and Eskimos of the Atlantic'*.

It is, of course, easier to lose, intentionally or by accident, stories handed down by word of mouth. Once they have gone, there is no one to oppose those who dominate with their own narrative. Then we forget why one thing is present in a landscape, why it used to be valuable, and what reasons we may have for looking after it.

## Concluding Comments

In this chapter, my aim has been to set the scene for a sustainable agricultural revolution by indicating that agricultural and food systems, and the landscapes they shape, are a common heritage to us all. For all our human history, we have been shaped by nature, while shaping it in return. In recent times, that shaping has been destructive, with food seen as a commodity and no longer part of culture. In our modern and industrial age, we are losing our languages, memories and stories about land and nature. These disconnections matter because they serve to promote a persistent dualism – that nature is separate from people, that nature can be conserved in wildernesses, and that economies can succeed without regard to the fundamental significance of agricultural and food systems.

## Chapter 2

# *Monoscapes*

## The Darker Side of the Landscape

The term 'landscape' first entered the English language from the Nether-lands in the 16th century, at the time when the Dutch were actively manipulating and redesigning their lands with new engineering methods for drainage. *Landschap*, like the German *landschaft*, meant both a place where people lived, as well as a pleasing object. Landscapes have inspired painters and poets in all cultures, and their designs have made many a view famous, even iconic. Great movements have emerged, and we celebrate beauty and perfection. Often the representations themselves have gained worldwide recognition, and so have entered cultures and become as important as the real landscape itself.

It is all too easy, though, to forget that landscapes themselves are also social constructions, with many different meanings bound up in them. A grassy hillock catches the eye and sets off the distant woods. To another viewer, though, the hill is a burial mound with ancient significance, or, worse, it hides the bodies of a recent war crime. A field of golden wheat stretches to a European horizon, and could yield 12 to 15 tonnes on every *hectare* in a good year. Yet, people still go hungry. The modern agricultural revolution of the second half of the 20th century transformed landscapes worldwide, and brought unprecedented levels of food production. World food production grew by 145 per cent during the 40 years to the year 2000, and even per person by 25 per cent, despite considerable population growth.[1] But as we all should know, this extraordinary 'success' still masks the persistent hunger of 800 million people.

Landscapes hide many ills, acts of unkindness and savagery perpetrated by people on other people. We look upon old landscapes with pleasure, and yet they can hide so much. Often they embody something deeply important to a whole culture: the dark, mysterious forests of central Europe, or the wide prairies and steppes of North America or Central Asia, or the spectacular rice terraces of Asian hillsides. Stephen Daniels and Denis Cosgrove say: *'a landscape is a cultural image'*, which implies both observation and separation, something with many codes and levels of understanding.[2]

The art historian John Barrell details some of the ambivalences in landscapes in *The Dark Side of the Landscape*.[3] The English pastoral landscape projects images of harmony with nature and of continuity; this was a foundation for Romantic notions of landscape. Yet, look closely, and the work of many painters raises questions about the relations between the land and people, and between people and people, particularly between the rich and poor. These questions apply widely. During the 18th and 19th centuries, many painters were directly commissioned by the wealthy, so it is hardly surprising that they should tend to present a partial construction of the landscape. Few painters depicted the country house and gentleman landowner in the same territory as the poor cottager. It is either the house in the landscape, or it is the labourer, hard at work, and somehow happy to be there. The labourers work continuously, and if they stop working, you suspect the vision might fade. As Barrell put it: *'it is not just that the rich have the power to be benevolent. . . but that the act of benevolence is an act of repression.'*

In these landscapes, there is paradox and tension. We are looking at cultural landscapes that are deeply rooted and persistent, or that at least come to embody timelessness. But the social aspects can imply a persistence of deeply rooted inequality and poverty. This is a good reason for believing that the conservation of a landscape without social change is only half of the picture. As we shall see later, all the recent significant progress with sustainable agriculture involves both social and natural transformations.

During this period, however, the idea of creating a harmonious and well-organized society was founded on continuous hard work, and labourers who, according to Barrell, *'do not step between us and the landscape —*

*they keep their place*. They are also obliged to feign a *'cheerfulness in adversity'*. Of course, there are clearly different interpretations. Some would say people are depicted as one with nature, while others point out that the people depicted do nothing but work, and would be disciplined if they stopped to gaze upon the view. The problem is that the pastoral and Romantic notions of landscape comprise a *'vision of rural life whereby the fruits of nature are easily come by more or less without effort'*, and this is clearly untrue.

According to Barrell, great artists such as Gainsborough, whether by accident or design, *'naturalize the extreme poverty of the poor — he presents it as a fixture in a changeless world which is the best of all worlds'*. Nonetheless, there is another important truth in these landscapes. Artists only worked with diversity, such as the big house, ruined abbey, or church framed with trees; the landowner and shadowed worker; the woodlands, meadows, cornfields and ploughed lands, pastures and meadows. Landscape art is nothing without diversity. It is the loss of natural diversity in the landscape that is one of the tragedies of modern industrialized agriculture.[4]

## Exclusions from the English Commons

The landscape itself is a type of common property. It can be enjoyed and appreciated if, of course, you are allowed to see it. The idea of commons implies connection, something people can enjoy either collectively or individually and from which they derive value. Over the centuries, two types of common management emerged in Europe. These were the common or open-field systems of cropland, which persisted for 1000 years, and the common management of wild resources, woodlands, pastures, wastes, rivers and coasts. In these systems, local people held rights for grazing, cutting peat for fuel (turbaries), cutting timber for housing (estovers), grazing acorns and beech mast (pannage), and fishing (piscary).

Over the years, however, both types of common came to be steadily enclosed and privatized, mostly as a result of the actions of landowners and the state, who were feverishly driven by the prevailing view that the commons were inefficient. The result was an extraordinary transformation of the landscape, particularly during the 18th and early 19th centuries. Local *enclosure* had occurred in the 17th century and earlier; but the process accelerated with the introduction of the parliamentary enclosure acts, dating from the early 18th century, which witnessed 2750 acts until 1845 – the date of the last general enclosure act. At the same time, 'wastes', heaths, moors and commons were enclosed through 1800 acts between 1760 and the 1840s.[5] Commissioners with extensive powers were appointed to redesign the landscape in more than 3000 parishes. As a result, 2.75 million hectares of common land were enclosed, comprising 1.82 million hectares of open-field arable, and 0.93 million hectares of 'wastes'. To put this in perspective, there are about 18 million hectares of agricultural land in the UK, of which just 4 million are currently under arable farming, and about 0.5 million still under common land.[6]

Historians have long documented the political and economic forces driving these enclosures, the powerful rhetoric used to support the claims for national progress, and the consequences for the wealthy and the poor. At the time, agricultural writers were unanimous about the agricultural benefits that derived from individual, as opposed to common, occupation of land. Most ignored the social losses caused by enclosures, and magnified the economic waste of the common use of land – both arable and 'wastes'. In his famous book *English Farming*, Lord Ernle records the views of dozens of notable writers of the 16th and 17th centuries, including Fitzherbert, Hartlib, Houghton, Lee, Moore, Norden, Taylor and Tusser, all of whom considered enclosure 'lawful' and 'laudable', and the commons wretched and wasteful.[7]

The narrative of the time was uncompromising. Silvanous Taylor said: *'this poverty is due to God's displeasure at the idleness of the commoners'.* From the pulpit, the Reverend Joseph Lee opined that the commoners fostered laziness, and Adam Moore said that the commons were overstocked, and were *'pest houses of disease for cattle. Hither come the poor, the blinde, lame, tired, scabbed, mangie, rotten, murrainous'.* John Norden was equally one-sided, saying that those who lived on wastes and commons were *'people given to little or no labour, living very hardly with oaten brew and sour whey. . . as ignorant of any civil source of life as the very savages among the infidels, in a manner which is lamentable and fit to be reformed'.* Despite these dominant views, it seems extraordinary that one short piece of folklore verse should have persisted to this day, as it seems to suggest a deeper truth: *'The law locks up the man or woman, Who steals the goose from off the common, But leaves the greater villain loose, Who steals the common from the goose.'*

Some writers did concede *'economic gain might involve social and moral loss'.*[8] A few activists even defended the rights of commoners, and movements for wider change arose, including those seeking to claim tracts of land for the public at large. Jerrard Winstanley and friends tried to establish a new common society in 1649 by settling on lands near Walton-on-Thames. They interpreted the defeat of Charles I in the Civil War as implying new rights for people to own their own land, and to use common resources. But they were mistaken, and Lord Fairfax's soldiers burned their huts and threw them off. Much later, William Cobbett, writing during his *Rural Rides* of the 1820s, noted something important about poverty, landscape and access to resources. Of the *monoscape* arable lands, he said:

*There were no hedges, no ditches, no commons, no grassy lines. . . and the wretched labourer has not a stick of wood, and has not a place for a pig or cow to graze. What a difference there is between the faces you see here, and the round, red faces you see in the wealds and forests.*[9]

During the late 17th, 18th and early 19th centuries, there was, of course, a period of extraordinary innovation in agriculture in Europe – so much so that this is now known as the Agricultural Revolution, as if it were the only one, rather than just the latest before our modern period. Over a period of about 150 years, crop and livestock production in the UK increased three to fourfold, as innovative technologies, such as the seed drill, novel crops such as turnips and legumes, fertilization methods, rotation patterns, selective livestock breeding, drainage, and irrigation, were developed by farmers and spread to others through tours, open-days, farmer groups, and publications, and then adapted to local conditions by rigorous experimentation.[10] However, throughout this time, the 'wastes' were never more than a symbol of backwardness. Arthur Young, great innovator, reformer and writer, was moved to call those who opposed enclosure 'goths and vandals', and as assistant tithe commissioner, he indicated that the heaths of Suffolk were *'mere sand encumbered with furze (gorse) and fit for nothing but rabbits and sheepwalk'*. After enclosures, poor farmers had to sell their animals, as they had lost rights to fodder beyond their farms; many, given smaller plots in lieu of grazing rights, sold their land and, according to Jane Humphries, *'the money was drunk in the ale house'*.[11]

The poet John Clare was an exception when he wrote with feeling about what had been lost. Most contemporary commentators focused on the economic gains from enclosure. He, by contrast, mourned the loss of memories accumulated over the ages, the open field system having persisted for 700 years by this time. In his journal, Clare wrote in 1824 about what had been lost:

*Took a walk in the fields and saw an old wood stile taken away from a favourite spot which it had occupied all my life. . . it hurt me to see it was gone for my affections claim a friendship with such things, but nothing is lasting in this world. Last year, Langley bush was destroyed, an old whitethorn that had stood for more than a century, full of fame. The gypsies, shepherds and herdsmen all had their tales of its history, and it will be long ere its memory is forgotten.*[12]

Not only are both the stile and the old named tree lost, but the memories, too. They persist for a while, perhaps for generations; but without renewal, they eventually die. The enclosures disenfranchised small farmers and commoners, and forced many to move to urban centres for work. So started the large-scale disconnection between people and the land, a process that continues today.

## Winners and Losers in the Wetlands and Forests

The story of the drainage of the low-lying *fens* of East Anglia illustrates how quickly some people became winners and others losers. The first major drainage of marshes for agricultural improvement occurred during the reigns of Henry VIII and Elizabeth I; but it was not until the 17th century that serious attempts were made on the Great Level of the Fens, a vast wetland of 280,000 hectares ranging across six counties of eastern England. Local people were hunters and gatherers, *'travelling in punts, walking on stilts, and living mainly by fishing, cutting willows, keeping geese, and wildfowling'*. But the official narrative of the time was that these areas comprised *'water putrid and muddy, full of loathsome vermine, the Earth spuing, unfast and boggie'*, and that these unproductive wetlands were conveniently *'overmuch harbour to a rude and almost barbarous sort of lazy and beggarly people'*.[13]

In the early 17th century, commissioners were appointed by government and backed by new legislation to speed the process of drainage. Cornelius Vermuyden, popularly accredited with bringing drainage know-how from the Netherlands to England, was appointed with the Earl of Bedford to lead the undertaking. Despite decades of technical and social setbacks, by 1649 a new system of drains, raised riverbeds, outfalls, sluices and dams was complete. Vermuyden reported that on this newly privatized land, *'wheat and other grains, besides innumerable quantities of sheep, cattle and other stock were raised, where never had any before'*. But it was not so simple, as these improvements provoked commoners and fen men to half a century of uprisings. They broke embankments, fired mills and filled drains. In some cases, they secured concessions. Ernle indicates that it was not until 1714 that the riots caused by the reclamations ceased. Yet, these protests were to no avail, as the fens stayed drained and in private hands.

Soon after this period, there followed one of the most notorious examples of state disenfranchisement of people relying on the resources of the commons. This was the passing, in May 1723, of the Waltham Black Act, or just 'Black Act', by the English parliament. In his compelling account, the historian E P Thompson describes how those in power took to new extremes their justification for wresting control of forests.[14] The act described the 'Blacks' as *'wicked and evil-disposed men going in disguise'* to pillage the royal forests of deer and do battle with forest officers. Critically, the Black Act created 50 new capital offences, which were then extended by successive legal judgements. Anyone found with their face *'black'*, or who might 'appear in any forest, close, park, or in any warren, or on any high road, heath, common or down', was now likely to be charged with a capital offence. Thompson quotes Sir Leon Radzinowicz's mid 20th-century judgement: 'It is very doubtful whether any other country possessed a criminal code with anything like so many capital provisions as there were in this single statute.'

The narrative of the time was, again, that commoners were destroying woods, coppices and heaths, and deliberately stealing the resources of others, particularly deer, game and fish. This made them, of course, poachers, smugglers and criminals, rather than simply rural people trying to make a living. What do the records tell us about these people who were caught and put to death? They were labourers, servants, millers, innkeepers, yeoman farmers, blacksmiths, butchers, carpenters, gardeners, ostlers, tailors, shoemakers and wheelwrights. They were 'again and again. . . men with small freehold or copyhold farms, sometimes scattered in several parcels in more than one parish, adjoining the heath and forest with their valued grazing and common rights'.[15] Not surprisingly, none were gentlemen farmers or squires. E P Thompson describes the act as 'savage' and 'atrocious'. For most of the 18th century, though, it directed and strengthened the majority of people's attitudes not only to common resources, but also to the people who relied upon them for their livelihoods. It also, because of Britain's rapidly growing empire, helped to shape lands and thinking in many other parts of the world.

## Commons and Exclusions in India

Enclosures of common pastures, swamps and grazing grounds have provoked exclusion and conflict in many other parts of the world.[16] Even though these are well documented by historians, today we are still doing more of the same, sometimes in the name of conservation, more often in the name of creating more productive farming. Very often, new social conflicts have come to threaten the success of the new system.

Madhav Gadgil and Ramachandra Guha's perceptive analysis of Indian ecological history, *This Fissured Land*, highlights the essential interdependence of ecological and social change. This is important because few histories have focused upon this vital connection between nature and people. As the authors indicate of India:

*A whole range of resources, regulated and utilised in many different ways, is under great stress. There are very few deer and antelope left to hunt for hunter-gatherers. . .*
*A majority of shepherds in peninsular India have given up keeping sheep for want of pasture to graze them. The shifting cultivators of north-east India have drastically*

*shortened their fallow periods. . . All over, peasants have been forced to burn dung in their hearths for want of fuelwood, while there is insufficient manure in fields. Groundwater levels are rapidly going down.*[17]

In recent decades, *common property resources* have been in steep decline in India, even though they form a significant part of rural people's livelihoods. As elsewhere, they have been neglected, over-exploited and privatized, and to all but the poorest are often invisible. N S Jodha's 30-year study of dryland villages illustrates just how drastic has been the change in community pastures, forests and *watersheds*, community threshing grounds, village ponds and rivers. He found that the poorest rely on common resources the most, as these annually provide up to 200 days of employment for each household, about one fifth of total income and four-fifths of all fuel and animal feed. But for the most wealthy, they rarely provide more than 2 per cent of income. In drought years, commons are even more important, when the poorest derive 40 to 60 per cent of income from these resources.[18]

Tony Beck and Cathy Naismith have put a monetary value on these common property resources, calculating that they contribute US$5 billion per year to the incomes of the rural poor in India, worth about US$200 per household. Following Jodha's groundbreaking study, further research has confirmed the fundamental value of these resources to rural people, and particularly to the poorest. These studies indicate that the commons contribute 12–25 per cent to rural livelihoods, and that the proportion is greatest for the poorest households – women and children are especially dependent upon them. They also confirm that the area and status of common property regimes have declined steadily over the past 50 years, as rights have been gradually removed and local institutions undermined.[19]

In Jodha's villages, the area of commons has fallen by 40–55 per cent per village since the 1950s. With population growth, this means that the number of people relying on each hectare of common has increased threefold. The sad truth is that these changes have been accompanied by a collapse in traditional collective management. Over this period, the number of villages with locally established regulations for rotational grazing, seasonal restrictions and provision of watchmen fell from 80 villages to just 8. Transgressors of these norms and regulations were formerly taxed, levied or fined in 55 villages; by the 1980s, it occurred in none. Users' social obligations to invest in the collective upkeep of watering points and fencing fell from 73 to just 12 villages.

It was once different. Gadgil and Guha tell us how pre-colonial kingdoms in India set aside elephant forests and hunting preserves, and how religion played a role in designing social mechanisms and obligations that promoted careful use of natural resources. They quote a third-century edict, in what is now Orissa, which stated that:

*Medical attendance should be made available to both man and animal; the medicinal herbs, the fruit trees, the roots and tubers, are to be transplanted to those places where they are not presently available, after being collected from those places where they usually grow. Wells should be dug and shadowy trees should be planted by the roadside for enjoyment both by man and animal.* [20]

Over time, communities developed locally specific regulations and rules for the care of natural resources. Often, named families were the forest guards; elsewhere, others would do all the harvesting and delivery of wood to households. Rules on hunting were common, such as the release of trapped pregnant does or young deer. These community regulations came under serious pressure during the colonial era. Timber was exported to Europe and used as sleepers in the expansion of the domestic railway network. Whole forests in the Himalayas were *'felled even to destruction'*, and hills in southern India *'to a considerable degree laid bare'*.

Wild common property resources are still important to many rural people in developing countries. The poorest are the most dependent upon the commons and are, of course, the least likely to have political power. Therefore, they are unable to prevent the loss or appropriation of these commons. Many have argued that commons are tragedies because they cannot be productive – too many collective constraints on the whole, too many free riders. Large-scale privatization, or enclosure, has been the result. This is no surprise, perhaps; but whether in England during the 18th century, or India during the later 20th century, the losers were always the poorest. In some cases, this was the intention; in others, it was an accidental but inevitable outcome. During enclosures, those with rights to commons were often bought off, and the money spent or the land repossessed. These histories of dispossession are long, deep and painful. Sadly, they persist today in the names of both conservation and agricultural modernization. [21]

## The Loss of Commons Knowledge in South-East Asia

The rice fields of South-East Asia are one of the wonders of the world. On a bright day, the azure blue of the watery fields sparkles, as snow-white egrets drift gently across the landscape. When grey clouds bring down the sky, the landscape takes on a moody presence. Where the hillsides are steep, terraced fields cut into the slope with extraordinary precision, like so many layers of a cake. It takes deep understanding to bend these landscapes and the water to the collective will. No one is quite sure when these methods of farming arose. In Bali, the first records of irrigated rice cultivation date to AD 882; since then, landscape management on a heroic scale has been built into the egalitarian Balinese *sawah* rice system.[22]

Irrigation cooperatives, the *subaks*, were responsible for the allocation of water and the maintenance of irrigation networks because wet rice farming is too complex for one farmer to practise alone. Each *subak* member had one vote regardless of the size of the landholding. Soil fertility was maintained by the use of ash, organic matter and manures; rotations and staggered planting of crops controlled pests and diseases; and bamboo poles, wind-driven noise-makers, flags and streamers scared birds. Rice was harvested in groups, stored in barns and traded only as needs arose. The system was sustainable for more than 1000 years. Yet, in the blink of an eye, rice modernization during the 1960s and 1970s shattered these social and ecological relationships by substituting pesticides for predators, fertilizers for cattle and traditional land management, tractors for local labour groups, and government decisions for local ones.

The benefit was this: modern rice varieties yielded 50 per cent more, though only under optimum conditions – the new rice was more susceptible to climatic and hydrological variations. Pests and diseases increased as a result of the continuous cropping and the elimination of predatory fish and frogs by pesticides. Farmers sold cattle, as they were no longer needed for ploughing and manures; mechanized rice mills displaced groups of women who used to thresh and mill the rice. Modern rice had to be sold immediately after harvest when the prices are low. This meant that men received large sums of cash, and women could no longer plan for the year's *food security* by monitoring the rice barn. The democratic *subak* organizations, once in complete control, lost decisions to government institutions, which decided cropping patterns, planting dates and irrigation investments. The reduced employment in rice cultivation forced rural people to seek work elsewhere; with the undermining of the *subaks*, goods were no longer redistributed from the better-off to the poorest through religious rituals.

The Indonesian and Malaysian islands and peninsula are also home to another remarkable cultural system called *adat*. This comprises more than indigenous knowledge or beliefs, more than a legal system. It is, as Patrick Segundad of the Kadazan community in Sabah says:

*. . .an unwritten understanding of common things that everybody should know. Adat is not only important in how we deal with our resources but also in how we live. It isn't like the concept of managing but rather that two things happen in the same time. While you might manage something, what you manage is also managing you. A person is part of a greater single action, a larger balance or harmony.*

This is the key. *Adat* shapes people's interactions with nature; people, in turn, are shaped by everything around them. Salfarina Gapor recently completed a study of the Melanau people in the coastal regions of Sarawak. Here, *adat* means harmony between spirits, humans, animals and plants, and it dictates social systems of joint bearing of burdens, reciprocal assistance and an ethic that protects the land and species biodiversity. The main staple for the Melanau is sago palm, and *adat* dictates a finely tuned set of management strategies. The contrast with modern rice methods elsewhere in the region is stark. The main predators of sago are monkey, boar and termites. At planting time, farmers plant just one palm sucker in the cleared field, and surround it with a variety of plants that are variously itchy, bitter and poisonous. They leave the field for three days to allow monkeys and boar the chance to come to the sago, and learn that it is not tasty. Farmers then plant out the rest of the sago, and the monkeys now eat the pests rather than the crop. But Gapor also found that *adat* is under severe pressure. Many young people do not know about it, and modern agricultural and plantation methods do not account for its sensitive understanding of ecosystems. The worry is that cultural and ecological knowledge are under threat.[23]

## Modern Dispossessions

Every continent has its own tragic histories of the dispossession of those who treat the land, or parts of it, as a common resource. The dark side of the world's first national park at Yellowstone is that Crow and Shoshone Native Americans were driven out of their lands by the US army, who then managed the park themselves for 44 years. Today, similar exclusions persist in parks that are constituted as strictly protected areas. The assumption that the conservation of natural resources is only possible through the exclusion of local people is pervasive throughout history. Local mismanagement has been used as an excuse to exclude people who may be in different tribes or who move about, rather than engage in settled farming. States adopt a variety of value-laden terms, such as scheduled tribes in India, minority nationalities in China, cultural minorities in the

Philippines, isolated and alien peoples of Indonesia, aboriginal tribes of Taiwan, natives of Borneo, and aborigines of Peninsular Malaysia. As Nancy Lee Peluso indicates: *'the terms are politicized by their application to particular users rather than uses'.*[24]

Many of these people, nomads, pastoralists, slash-and-burn hill tribes, hunter-gatherers, gypsies, and itinerants, have been a thorn in the sides of states.[25] States have tried to settle them, or have moved settled peoples into their regions, such as the massive forced 'transmigrasi' of Javanese rice families to the outer regions of Indonesia during the 1980s. Excluded from their own rice cultures and landscapes, the Javanese were resettled to new areas that were inappropriate for rice cultivation, and which were already full of local people. Conflicts were inevitable, and neither dispossessed group benefited. These changes echo the experience of transportees who were excluded from Britain during the 18th and 19th centuries, many for minor misdemeanours after the loss of their lands during the enclosure. These transportees were relocated to Australia, where they took over land from Aboriginal tribes.

This forced resettlement is deeply damaging to people. Kaichela Dipera, a Mukalahari from Botswana, says of the Bushmen of the Kalahari Game Reserve:

*The experience of moving away is so painful when you think of it because they are moving from a place where they have been living for a long time. They know what the plants are for; they know the source of water and food. When people are moved to a new place they are cut off entirely from their culture and are moved to a place where they must start a new culture.*[26]

In truth, such disconnections are more than painful. They take away people's sense of the meaning of life, and the memories of dispossession can last for generations.

The savannahs of East Africa are world renowned for their wildlife. Yet, they have emerged as a result of a long process of co-evolution between pastoralists, their cattle, and local wildlife. Without one, the others suffer. When the Maasai were expelled from their lands in Kenya, the newly created parks were colonized by regenerating scrub and woodland, leaving less grazing for antelopes.[27] Even greater harm is caused by agricultural development. One of the most notorious cases comes from Tanzania, where wheat farms were imposed on the dry Basotu Plains from the late 1960s to the early 1990s. These plains are the homeland of more than 30,000 Barabaig pastoralists, whose culture is based upon the keeping of livestock and common use of forage, water and salt resources scattered throughout their territory. A complex grazing rotation system with eight forage regimes means that some land is free of people and animals for long periods, thereby preserving it from overuse. All members of the community have access to communal land, which is protected by customary rights and obligations for individuals, clans and local groups. The Barabaig, like many people who live in harsh environments, have a tradition of respect for the land on which they rely for their survival. Their elders say: *'We value and respect the land. We want to preserve it for all time.'*

But in order for wheat to be grown on the Basotu Plains, about 40,000 hectares of the most fertile land was taken from the Barabaig. For a few years, these farms came to supply half of the national demand for wheat. A narrowly focused project evaluation arrived at a positive cost-benefit ratio, and the nearly 40 per cent return to invested capital indicated that it was a *'very profitable investment for the Tanzanian economy'*. But if the wider social and environmental impacts had been counted, then a very different picture would have emerged. Charles Lane spent several years documenting first hand the severe impact upon local people. Although the wheat farms covered only one eighth of their land, this was their best grazing land, and the loss was crucial. By losing access to the most fertile areas, the whole rotational grazing system was compromised, resulting in a drastic reduction of livestock numbers. Many of their sacred graves were ploughed up, and as the soil was left bare after harvest, so erosion silted up the sacred Lake Basotu. The problem was that outsiders fundamentally misunderstood the pastoralists and their strategies for managing common rangeland. Herders move in response to their assessment of range productivity, and those who fail to understand this can be misled into thinking that land is vacant or poorly managed. One study said: *'The project has many of the characteristics of a frontier development effort. Traditional pastoralists. . . are being displaced and absorbed into the project as labourers. Previously idle land is being brought under cultivation'.* The project has now closed, but the effects on local people remain.[28]

## Forest Rights and Protection in India

Concerns about the destruction of nature in India were formalized by national policy-makers in 1864 with the establishment of the Imperial Forest Department, and a year later with the first Indian Forest Act. This marked the steady extension of state control over forests that would continue unabated until the early 1990s, when the idea of joint forest management was given policy support. During the 19th century, forests were under pressure, largely from the imperial power itself. It took control and added, over time, a narrative about local people's inability to manage these resources with care. The Forest Act had no provisions for defining local customary rights to exploit natural resources – a blind spot that persisted for more than a century.

Administrators did, however, distinguish between rights that were not permitted, and privileges that were granted as concessions to graze and collect firewood. Gadgil and Guha indicate that such privileges were *granted by the policy of the government for the convenience of the people*.[29] In practice, there were many interpretations, from those who argued the state should annex and take complete control of all forests, to administrators who argued that where customary use existed, it should also be granted legal rights. According to Gadgil and Guha, the first inspector general, Dietrich Brandes, was a pragmatist who drew comparison between common rights to the New Forest in England and indigenous management of forests in India. He advocated the restricted take-over of forests by the state, and wrote appreciatively of common regulation of forests and the extended network of sacred forest groves.

However, Brandes lost out, and the subsequent 1878 Indian Forest Act set the scene for another century by granting forests and punitive sanctions to the forest department. Between 1878 and 1900, the area of designated state forests grew from 36,000 to 200,000 square kilometres, of which 40 per cent comprised protected forests. By independence, the total had grown to 250,000 square kilometres. Meanwhile, the forest department evolved into a revenue-raising department, rather than a resource manager, and its success was judged on income rather than the stock of biodiversity maintained. Predictably, this marginalized those who depended upon wild resources, such as hunter-gatherers, shifting agriculturalists and settled farmers and artisans who relied upon forest products for house construction, basket-making, musical instruments, furniture, weaving, tanning and dyeing.[30]

An inevitable result of such exclusions and denials of rights is that local people are forced to struggle for their land. Over the last two decades of the 20th century, the expansion of national parks and protected areas which permitted no, or very limited, use of local resources continued at a rate of 600,000 hectares per year, resulting in the forcible displacement of many thousands of people. This has provoked many open protests, rallies and acts of sabotage against national parks and protected areas themselves. In the early 1980s, more than 100 clashes were reported from national parks and sanctuaries in India. Later, villagers set fire to large areas of the Kanha and Nagarhole National Parks during the early 1990s, when denied access to the park for forest products. In remote areas, insurgents have taken advantage of local resentment to take over a tiger reserve in Assam and drive out forest guards, and to invade a tiger and buffalo reserve in Madya Pradesh, where 52 villages of tribals had been evicted.[31]

Enlightened professionals realize that imposed modes of conservation simply do not work. They are expensive – much of the budget for strictly protected areas has to be spent on aircraft, radios, weapons, vehicles, salaries of armed guards, night goggles and other 'anti-poaching' equipment.[32] They are also often ecologically counter-productive. In the Keoladeo Ghana National Park in Rajasthan, the Bharatpur wetlands support many birds, including wintering geese, ducks and the endangered Siberian crane. A ban on buffalo grazing, established in 1982 for the cause of crane conservation, provoked violent conflict between local people and the police, resulting in several deaths. The ban was reinforced, but paspalum grass began to grow unchecked, choking the water bodies and making the habitat unsuitable for water birds. Money had then to be spent on bulldozers to remove the grass; but this was never as efficient as buffalo grazing. There has been some recent progress – though only, as Madhav Gadgil put it, to the point that *villagers are now allowed to harvest the grass by hand*.[33]

Such local concerns led to the establishment, during the 1970s, of the Chipko movement, now one of the most famous of environmental movements. It began when local people in the Himalayas were refused permission to fell their own trees in the Alakananda valley. The government then allocated the same forest to a distant sports-goods firm for their sole use. Chipko means 'to hug' in Hindi, and villagers did exactly this to trees that they wished to protect. The idea was compelling and simple, and it spread quickly through Uttar Pradesh, and eventually to southern India, where it came to be known as the Appiko movement ('to hug' in Kannada). Importantly, these were both environmental and social movements. They made the point that people cared, and they would do something about it. It was from this movement that the idea of joint forest management emerged, which received official government support during the early 1990s. Evidence had shown that if people are given responsibility for their natural resources, they can be effective at both increasing productivity and ensuring that the benefits are shared. Handing over such rights does not mean the tragic destruction of forest resources.

## Saving Nature in Protected Areas and National Parks

The idea of the wildernesses is compelling, and it forms a central part of the writings of John Muir, known by many as the father of conservation. Claimed by both Scotland, where he was born, and by the US, where he lived until his death in 1914, Muir's writings and campaigns gave rise to the world's first national parks. He helped to found the Sierra Club in 1892, an environmental movement with 600,000 members today. Many commentators talk at length about the wilderness Muir frequented. In 1869, he walked the Sierra Nevada Mountains, and lived rough for five years to study the flora, fauna and geology. Muir accompanied shepherds with their flock of several thousand sheep from the foothills to the high mountains, including the headwaters of the Merced and Tuolumne rivers and the spectacular waterfalls of Yosemite Creek. He called Yosemite a 'park valley', and celebrated nature's creativity: *'what pains are taken to help this wilderness in health. . . How fine Nature's methods! How deeply with beauty is beauty overlaid'.*[34]

Yet, this is a landscape shaped by humans, and in particular by the Ahwahneechee, who created the meadows of Yosemite through fire clearances.[35] Muir was aware of the effects of people on the landscape – he carefully documented the actions of the shepherds and local Native Americans whom he met on the way. But this awareness is lost on many commentators, who themselves see only untouched wilderness through Muir's eyes. He encountered groves of Sabine pines, the nuts of which, he was told by a shepherd, were gathered by the 'Digger Indians' for food. These groves were not there by accident; they had been sustained and protected by the gatherers. Muir observed women collecting wild lupin, saxifrage and roots, and recorded a variety of other species as valuable food sources, including beaked hazel nuts and acorns, squirrels and rabbits, berries, grasshoppers, black ants, wasps, bee larvae, and many other *'starchy roots, seeds and bark in abundance'*. At one stage, in early July, Muir and his colleagues ran out of food, apart from mutton. Awaiting supplies amidst gnawing hunger, Muir lamented the fact that they could not find food in this rich landscape: *'Like the Indians, we ought to know how to get the starch out of fern and saxifrage stalks, lily bulbs, pine bark etc. Our education has been sadly neglected for many generations.'*

Muir noted the soft touch of the Native Americans on the landscape:

> *How many centuries Indians have roamed these woods nobody knows, probably a great many. . . and it seems strange that heavier marks have not been made. Indians walk softly and hurt the landscape hardly more than the birds and squirrels. . . How different are most of those of the white man, especially on the lower gold region — roads blasted in the solid rock, wild streams dammed and tamed and turned out of their channels.*

He also noted that the Native Americans created *'enchanting monuments. . . wrought in the forests by the fires they made to improve their hunting grounds'*. It would, therefore, be churlish to be too critical of Muir, or indeed of other wilderness writers such as Thoreau, as they were on a mission to save the remnants of primaeval nature, which had intrinsic value and which was under growing threat. It was their inspired writings that captured the imagination of readers sufficiently to lead to the establishment of the world's first national parks in Yellowstone in 1872, and later in Yosemite in 1890.

The harm that has been done lies in the perpetuation of a notion that we are separate from nature. Ironically, this is the very *modernist* problem that these writers were trying to oppose. The argument goes like this: as nature is separate from us, so it should be strictly protected in pockets and patches away from human activity. If it is protected, then we can shrug our shoulders at damaging economic activity in the surrounding land-scape. This is enclave thinking, and it is a simplistic narrative: let the farming and food production occur in one place, and let it do as it wishes.[36] The more productive it is, the less pressure is put on wildernesses and parks. This dichotomy of thought and action is damaging, both to farming and to the preservation and conservation of nature. It is built on the idea that nature which exists on agricultural land is largely worthless. But what about the tens of millions of monarch butterflies that migrate across the American plains, to and from Mexico each year, or the bio-diversity that flourishes in urban gardens? It is also built upon the idea that wildernesses exist untouched and unshaped by humans, and should be maintained that way. This is a serious myth of disconnection. It has led to great damage. It is now time to rethink these connections.[37]

The world's first formal protected area was established on 1 March 1872, when US President Ulysses Grant designated 900,000 hectares of north-west Wyoming as the Yellowstone National Park. The next to appear was in 1885 when the state of New York set aside nearly 300,000 hectares of the Adirondacks as a forest preserve. In neither case was the conservation of nature and wilderness the primary goal. At Yellowstone, the aim was to limit private companies from acquiring the geysers and hot springs. In the east, New York City's concern was to maintain its water-sheds and drinking water supply. These protected areas were followed by the 1890 designation of Yosemite National Park, and the 1891 amendment to the act revising land laws that permitted the president to create more forest reserves (later named national forests). Subsequently, President Benjamin Harrison proclaimed 15 reserves over more than 5 million hectares. But reversals soon followed designations, such as the 1897 Forest Management Act that allowed reserves to be cleared for timber extraction. Such advances and reversals have continued to the present day.[38]

Over the past century, parks and nature reserves have become the primary means of conserving nature, both for wildlife and for whole landscapes. According to the *United Nations List of Protected Areas*, there were 12,754 official protected areas worldwide in 2001, covering an area of 13 million square kilometres, an area larger than Brazil, China or the US.[39] Until the end of the 1950s, United Nations (UN) listed sites were designated at a rate of 300–400 per decade; this rose during the 1960s to more than 1000; to 2500 during the 1970s; to 3800 during the 1980s; back to 1800 during the 1990s. The World Conservation Monitoring Centre records an additional 17,600 protected areas on its database that are smaller than the UN's 1000-hectare minimum criterion, adding another 28,500 square kilometres to the total. All 30,000 protected areas now account for 8.83 per cent of the world's land area. Of the 191 countries with protected areas, 36 contain 10–20 per cent of their territory as protected areas, and a further 24 have more than 20 per cent.[40]

Protected areas are divided into six types along a spectrum from strict protection, to sustainable management and use of resources. One third of all protected areas, numbering 10,700 and covering 7 million square kilometres, are in categories 1–3, permitting no local use of natural resources. Of the 7322 protected areas in developing countries, where many local people still require wild resources for some or all of their livelihoods, 25 per cent are strictly protected in Asia and the Pacific, 28 per cent in Africa, and 40 per cent in Latin America. Of the 13 million square kilometres in protected areas, 7 million are strictly protected – 46 per cent of which are in Africa, Asia and Latin America (see Table 2.1).[41] This is a huge area of land from which people are actively excluded. The problem, as Nancy Lee Peluso has put it, is that *'managed biodiversity is hardly discussed in the current fervour of concern over losses of biodiversity, even in habitats (such as mangroves) that have clearly been occupied by humans for decades or centuries'*.[42]

The concept underlying the designation of protected areas is the conservation of a 'natural' state untouched by people.[43] As Arturo Gómez-Pompa and Andrea Kaus put it, these areas are seen as *'pristine environments similar to those that existed before human interference, delicately balanced ecosystems that need to be preserved for our enjoyment and use'*. This is not to say that they do not work. A recent study of 93 national parks of 5000 hectares or more in size in 22 tropical countries has found that formal designations do protect biodiversity. All the studied parks were more than five years old and were subject to human pressure, with seven out of ten having people living within their boundaries. One half had residents who contested the government's ownership of some part of the park. Yet more than eight out of ten of the parks had as much vegetation cover as when they were established. Parks suffered less degradation than the

surrounding undesignated area, with policing appearing to help, particularly in stopping illegal logging. The most effective parks were those with
clearly marked boundaries and close and cordial relations between
authorities and local communities.[44]

Nature clearly existed perfectly well before humans intervened, and will
do so after we disappear. But for the moment, we must recognize that most
landscapes are fundamentally shaped by human imagination and action.
This is a continuous dance, a tight coupling of nature and humans, the
outcome of which is what we see around us every day. Equally, though,
we should not conclude that all nature is an emergent property of human–
environment connections.[45] Baird Callicott and colleagues suggest
a middle way for conservation biology. These polarities are helpful
metaphors and rhetorical devices in order to focus debate; but most people
in practice stand somewhere on a spectrum between extremes. There
are such things as wildernesses, and there is a need for protection and
controls. Most 'wild' nature, though, is an emergent property of human
interventions; globally, most biodiversity occurs in human-dominated
ecosystems. This means that human decisions and visions matter, as they
can make a difference by provoking all of us to think and act differently.
But do we have the desire to redesign this relationship? Can we, as if by
alchemy, imagine different outcomes?

## Modernism and Monoscapes

Around the time that Muir and Thoreau were writing about wildernesses,
Brandes was forming forest policy in India, the first national park was
being established in the US, the enclosures had finally ended in the UK,
and the Japanese were coming to the end of their Edo period. Edo was
the largest city in the world in the 19th century, with more than 1 million
people and a population density three times as great as today's Tokyo. For
close to three centuries, Edo gave rise to extraordinary artistic and cultural
innovation, producing all of the major Japanese art traditions – tea
ceremonies, flower arranging, Noh and Kabuki dramas, distinctive styles
of architecture, urban design and landscape painting.

According to architect Kisho Kurokawa, '*Edo was known as the city of
blossoms*' – a metaphor for innovation, but also for the greenness of the city's
parks and gardens. Kurokawa believes that one of the most important
features of Edo was the hybrid organic nature of design. Diversity was
good, and anything that worked could be used in urban or rural space.
This was not a recipe for chaos, because principles of harmony determined
what would work. But diversity meant synthesis, and the synergistic

process of bringing together different elements to create a whole more significant than the sum of its parts.

The simplicity of the Japanese tea-room tells us something important about how we might design on a landscape scale. The first thing to note is that tea-rooms are not designed, they emerge, *'built through a process of natural accretion'*. Tea masters used only locally available material and had an ability to discover beauty and harmony in commonplace objects, such as trees, fallen branches or decayed boards. The important point here is that these items have multiple meanings. Simple rough thatch, for example, is there to remind you of the splendour of cherry blossom in spring, as well as of the luminous red maple leaves in autumn. This is the ambiguous code for Edo, with simplicity and harmony producing a living and changing series of landscape symbols, in which diversity grows over time as the system responds to incremental changes that people make.

Japanese landscape painters were more willing than Western landscape painters of the time to adopt a variety of formats, such as very wide screens or tall parchments. Their landscapes were always diverse – harmonious green hills covered with clumps of pink flowering cherries, set against a golden mist. These hills are the *satoyama* of myth and mystery, deeply embedded in Japanese culture and part of a rural vision called *satochi*. These *satochi* are areas that are marked by great diversity in the relationships between humans and nature, and embody the ideas of a path to mutual compatibility for both nature and people. They contain *furusato* – old settlements, places of community which give a special feeling to people. Many, too, were commons, known as *iriaichi*, which persisted without ecological destruction until the mid 20th century. Overall they are culturally important *diverscapes* of paddy rice, orchard trees, groves, hills, rivers and high mountains. Today, though, *satochi* are under threat because of modern patterns of economic development.[46]

This is the problem: modernism creates monoscapes. It is a kind of fundamentalism because it suggests that there is only one way, and no others can be correct. Monoscapes are dysfunctional systems. They are good at one thing, but people do not much care for them. In truth, a monoscape is less valuable than it appears, largely because value is captured and claimed by a small number of *stakeholders*. Poverty persists in the monocultural ideal, though it is clearly present in many societies that we may wish to call traditional. There is also social injustice at the core of wilderness monocultures. In order to make them 'wild' and 'untouched', the people who live there have to be removed. They are then replaced by tourists, who visit to experience the natural and real landscape upon which a new order has been imposed. By contrast, a polycultural approach accepts differences and the value of the whole.

When I use the term modern to describe current agricultural systems, I mean it in the philosophical rather than temporal sense. Such systems are certainly modern because they are what we have now. But, more importantly, they are modernist because they are single-code systems. Kurukawa was also a designer of the 1981 Royal Academy exhibition in London on *The Art of the Edo Period*, and he says it perfectly:

> *I do not reject the modern by any means. . . But when I see how rigid it has become, how it has lost all flexibility, I am forced to ally myself with those who attack the weaknesses of the modernist doctrine.*

A modernist agriculture is single coded – it does one thing (produces food) and it does it well. It draws on no local traditions; it is placeless, inflexible and monocultural. Diverscapes, by contrast, have more elements, more connections between these elements, and thus greater potential for synergies.[47] The post-modern is more symbiotic, and according to Kurukawa, *'from the intermediary space between these opposing poles many creative possibilities will well up'*.[48]

Landscapes are commons; yet, today they are increasingly shaped by non-local and global interests. These commons can never respond to the particular needs of the local, nor be able to change direction rapidly when something goes wrong. The landscape commons have been appropriated to a vision of efficient 'mono-use' and 'mono-culture'. We have to find new ways in which to claim back these commons, and to step outside the conservation-production dualism.[49] Who tells the landscape stories matters, as does who constructs the visions. If it is the powerful, defining a vision for the landscape and putting up the money, then we will see one type of outcome. If it is many individuals and small groups developing genuinely radical visions, then we will get something very different.

## Repossessing Natural Places

The term landscape has come to mean a pictorial representation of the countryside. Paradoxically, though, timeless and cultural landscapes may be allowing deep inequalities to persist. Thus, a landscape conserved without social change is only half the picture. Transformations are needed in both the natural and social spheres, and in their interactions and connections. Ultimately, transformations are needed in the way we think.

Writing in the mid 19th century, Thoreau was worried about our destruction of nature, and why protecting, conserving and understanding it mattered. He is particularly famous for his public departure from the

town to live by Walden Pond in the forests of Massachusetts.[50] For 26 months, he repossessed his own nature. In his account of life in the forest, he compared his views with those still in the town, and explored the nature of civilization, the economic exploitation of nature, the simple life, and the distinct sounds and deep solitude. His is the celebration of nature as a special place, not as a strictly untouched wilderness: *I went to the woods because I wished to live deliberately, to front only the essential facts of life, and see if I could not learn what it had to teach, and not, when I came to die, discover that I had not lived.'*
His contemplation changed him:

> *Sometimes, in a summer morning. . . I sat in my sunny doorway from sunrise till noon, rapt in a reverie amidst the pines and hickories and sumachs. In undisturbed solitude and stillness, while the birds sang around or flitted noiselessly through the house. . . I grew in those seasons like corn in the night.*

He discovers an intimacy with nature through such close observation, and through farming his bean field: *'consider the intimate and curious acquaintance one makes with various kinds of weeds'.*

The real insight in Thoreau's writing is the journey he himself travels, and his vision and willingness to experiment, and his desire to make his words meaningful to other people in the cities to whom he does, of course, return. *'I learned this, at least, by my experiment: that if one advances confidently in the direction of his dreams, and endeavours to live the life which he has imagined, he will meet with a success unexpected in common hours'.* His concern is with how we live our lives, each of us, and how this can be improved through a closer relation with nature. More than a century later, wildness writer Barry Lopez makes a similar connection: *'As I travelled, I came to believe that people's desires and aspirations were as much a part of the land as the wind, solitary animals, and the bright fields of stone and tundra.'*[51]

Such pride in your own landscape is common the world over. David Arnold quotes the renowned Bengali poet and novelist Rabindranath Tagore who, writing in 1894, said:

> *Many people dismiss Bengal for being so flat, but for me the fields and rivers are sights to love. With the falling of evening the vault of the sky brims with tranquillity like a goblet of lapis lazuli, while the immobility of afternoon reminds me of the border of a golden sari wrapped around the whole world. Where is there another land to fill the mind so?*

But Tagore also knew of another painful truth, and he rightly points to the combined social and ecological challenge: *'every house has rheumatism, swollen legs, colds or fevers, or a malaria-ridden child ceaselessly crying [whom] no one can save'.*[52]

The idea of landscape redesign combined with such social challenges is appealing; but, in truth, few have taken a radical view of what can be achieved.[53] This is precisely why changes brought about by today's newly emerging sustainable agriculture revolution are so important. Cultural and natural landscapes are being transformed precisely because some power is being put in the hands of the poorest; they are alchemists bringing forth a new world. It is the desires, aspirations and stories of these individuals that we must harness for a new connection between people and nature. We are fortunate that so many heroes have recently found a way of meeting food needs, while not damaging nature. It can be done, but it is difficult. The path towards sustainability, which is taken by individuals in remote places that are far removed from industrialization, must be adopted by all of us.

I once stood upon the top of the Temple of the Giant Jaguar, 96 metres above the floor of Tikal, the long-since abandoned capital of the Mayan empire. Below were the crowns of giant rainforest trees, the branches of which cracked and snapped as howler and spider monkeys leapt and chattered. A storm swept across this Petén forest of Guatemala and lashed me with ferocious wind and rain. Later, I reversed my way down the vertical step ladders, and then to the dizzying steps of the lower slopes of the pyramid. Had I been dropped here from afar, I may have been forgiven for thinking that I gazed upon a wilderness. The Petén is, after all, one of the world's hot spots for biological diversity, containing 200 species of mammals and 500 species of birds. I would have been right to be awed, but wrong about the wilderness. During the Mayan Golden Age between AD 250–900, Tikal alone supported a population of 10,000 to 40,000 people.

Since the mystifying collapse of the Mayan civilization, indigenous people have farmed with slash-and-burn methods. Fields are cleared in the forest, cropped for a couple of years, and then abandoned as families move on to new sites. Over time, as the population has increased, and as others came to log the forests, so farmers have had to reduce their fallow periods. As a result, they returned to former fields too soon for natural soil fertility to have been restored. Both agriculture and the forest come under pressure – yields remain low or fall, and the forest steadily disappears.[54]

But on the edge of Petén forest, farmers are using a magic bean to improve their soils and to save the rainforest. Some decades ago, the velvetbean (*Mucuna pruriens*) was introduced to Central America, probably from South Asia via the US. It did not spread far until several Honduran and Guatemalan non-governmental organizations, in particular World Neighbors, Cosecha and Centro Maya, discovered during the 1980s and

1990s that its cultivation with maize substantially increased cereal yields. The important thing is that *Mucuna* is grown as a soil improver. It can fix 150 kilogrammes of nitrogen per hectare each year – a free resource for farmers. For every hectare, it also annually produces 50–100 tonnes of biomass. This plant material is allowed to fall on the soil as a green manure, suppressing weeds and helping to build the soil. In this bean lies the protection of the Petén rainforest. Build the health of the soil, and farmers no longer want to burn trees in order to create new fields. Reclaiming land for agriculture is, after all, difficult and dangerous work, and farmers would love an alternative. An improvement to soil health changes the way that farmers think and act. They see the benefit of staying in the same place, and of investing in the same fields for themselves and their children.

Seven years after first standing at Tikal with Sergio Ruano, I came back to see how far farmers had developed their new settled ways. I walked with another colleague from Centro Maya, Juan Carlos Moreira, near the Usumacinta River, the Guatemalan border with Mexico. It is another area of extraordinary biodiversity. Inside the forest – this silent and eerie natural cathedral – the air was heavy with humidity and pierced with sunlight let in through holes in the canopy far above. By some wonderful coincidence of names, this was called the Cooperativa La Felicidad, or Happiness Cooperative. On this real political and administrative frontier, 250 farmers now grow *Mucuna* in their fields and have begun a journey across a cognitive frontier, towards settled and sustainable agriculture. I asked one, Gabino Leiva, about the bean manure, as they call it: '*The bean manure destroys the weeds; the beans simply kill them, and all the crops flourish much more. This is what we all need to do – manure our soil for increased production.*' It is technically easy. Improve the soils through low-cost, environmentally sound methods, save the remaining rainforests, and reclaim the monoscape for the people who live there.

There remain, of course, many confounding factors. The forests may still disappear under the chain-saws of the loggers; farm families still lack access to markets; and the adoption of these new settled systems of agriculture necessarily means the loss of systems of shifting agriculture, with their associated knowledge and sub-climax biodiversity. Moreover, this progress towards sustainable agriculture is being made despite current policies. What would happen if we were able to get these right, too?

## Concluding Comments

In this chapter, I began with some reflections on the darker side of the landscape. Throughout history, there are painful stories of exclusion, with the poorest and powerless removed from the very places and resources on which they rely for their livelihoods. It is easy to miss these exclusions, as they are wrapped in picturesque representations of landscapes, combined with a narrative of inevitable economic progress. Exclusions have arisen from both modern agricultural development and from the establishment of protected areas – both of which simply disconnect people from the nature they value and need. One third of all protected areas, covering some 7 million square kilometres, permit no use of resources by local people. Repossession of natural places is now a priority, and there is progress on a small scale. Systemic change, however, will need the collective actions of whole communities with access to the appropriate technologies and knowledge, and supported by appropriate national and international policies.

## Chapter 3

# *Reality Cheques*

## The Real Costs of Food

When we buy or bake our daily bread, do we ever wonder how much it really costs? We like it when our food is cheap, and complain when prices rise. Indeed, riots over food prices date back at least to Roman times. Governments have long since intervened to keep food cheap in the shops, and tell us that policies designed to do exactly this are succeeding. In most industrialized countries, the proportion of the average household budget spent on food has been declining in recent decades. Food is getting cheaper relative to other goods, and many believe that this must benefit everyone since we all need to eat food. But we have come to believe a damaging myth. Food is not cheap. It only appears cheap in the shop because we are not encouraged to think of the hidden costs, in terms of damage to the environment and to human health as a result of agricultural production. Thus, we actually pay three times for our food – once at the till in the shop, a second time through taxes that are used to subsidize farmers or support agricultural development, and a third time to clean up the environmental and health side effects. Food looks cheap because we count these costs elsewhere in society. As economists put it, the real costs are not internalized in prices.[1]

This is not to say that prices in the shop should rise, as this would penalize the poor over the wealthy. Using taxes to raise money to support agricultural development is also potentially progressive, as the rich pay proportionally more in taxes, and the poor, who spend proportionally more of their budget on food, benefit if prices stay low. But this idea of fairness falters when set against the massive distortions brought about by modern agricultural systems that, additionally, impose large environmental and health costs throughout economies. Other people and institutions pay these costs, and this is both unfair and inefficient. If we could add up the real costs of producing food, we would find that modern industrialized systems of production perform poorly in comparison with sustainable systems. This is because we permit cost-shifting – the costs of ill-health, lost biodiversity and water pollution are transferred away from farmers, and therefore are not paid by those producing the food, or are included in the price of the products sold. Until recently, though, we have lacked the methods to put a price on these side effects.

When we conceive of agriculture as more than simply a food factory – indeed, as a *multifunctional* activity with many side effects, then this idea that farmers do only one thing must change. Of course, it was not always like this. Modern agriculture has brought a narrow view of farming, and it has led us to crisis. The rural environment in industrialized countries suffers, the food we eat is as likely to do as much harm as good, and we still think that food is cheap. The following words were written more than 50 years ago, just before the advent of modern industrialized farming:

*Why is there so much controversy about Britain's agricultural policy, and why are farmers so disturbed about the future?. . . After the last war, the people of these islands were anxious to establish food production on a secure basis, yet, in spite of public good will, the farming industry has been through a period of insecurity and chaotic conditions.*

These are the opening words to a national enquiry that could have been written about a contemporary crisis. Yet they are by Lord Astor, written in 1945 to introduce the Astor and Rowntree review of agriculture. This enquiry was critical of the replacement of mixed methods with standardized farming. The authors insisted that: '*to farm properly you have got to maintain soil fertility; to maintain soil fertility you need a mixed farming system*'. They believed that farming would only succeed if it maintained the health of the whole system, beginning, in particular, with the maintenance of soil fertility: *'Obviously it is not only sound business practice but plain common sense to take steps to maintain the health and fertility of soil.'*[2]

But during the enquiry, some witnesses disagreed, and called for a *'specialized and mechanized farming'* – though, interestingly, the farming establishment at the time largely supported the idea of mixed farming. But in the end, the desire for public subsidies to encourage increases in food production took precedence, and these were more easily applied to simplified systems, rather than mixed ones. The 1947 Agriculture Act was the outcome, a giant leap forward for modern, simplified agriculture, and a large step away from farming that valued nature's assets. Sir George Stapledon, a British scientist knighted for his research on grasslands, was another perceptive individual well ahead of his time. He, too, was against monocultures and was in favour of diversity, arguing in 1941 that *'senseless systems of monoculture designed to produce food and other crops at the cheapest possible cost have rendered waste literally millions of acres of once fertile or potentially fertile country'.*[3] In his final years, just a decade after the 1947 act, he said:

> *Today technology has begun to run riot and amazingly enough perhaps nowhere more so than on the most productive farms. . . Man is putting all his money on narrow specialisation and on the newly dawned age of technology has backed a wild horse which given its head is bound to get out of control.*

These are wise words from eminent politicians and scientists. But they were lost on the altar of progress until now, perhaps – as new ideas on agriculture have begun to emerge and gather credence.

## Agriculture's Unique Multifunctionality

We should all ask: what is farming for? Clearly, in the first instance, farming produces food, and we have become very good at it. Farming has become a great success, but only if our measures of efficiency are narrow. Agriculture is unique as an economic sector. It does more than just produce food, fibre, oil and timber. It has a profound impact upon many aspects of local, national and global economies and ecosystems. These impacts can be either positive or negative. The negative ones are worrying. Pesticides and nutrients that leach from farms have to be removed from drinking water, and these costs are paid by water consumers, not by the polluters. The polluters, therefore, benefit by not paying to clean up the mess they have created, and they have no incentive to change their behaviour. What also makes agriculture unique is that it affects the very assets on which it relies for success. Agricultural systems at all levels rely for their success on the value of services that flow from the total stock of assets that they control, and five types of assets (natural, social, human, physical and financial capital) are now recognized as being important.[4]

*Natural capital* produces nature's goods and services, and comprises food (both farmed and harvested or caught from the wild), wood and fibre; water supply and regulation; treatment, assimilation and decomposition of wastes; nutrient cycling and fixation; soil formation; biological control of pests; climate regulation; wildlife habitats; storm protection and flood control; *carbon sequestration*; pollination; and recreation and leisure. *Social capital* yields a flow of mutually beneficial collective action that contributes to the cohesiveness of people in their societies. The social assets that comprise social capital include norms, values and attitudes that predispose people to cooperate; relations of trust, reciprocity and obligations; and common rules and sanctions that are mutually agreed upon or handed down. These are connected and structured in networks and groups.

*Human capital* is the total capability that resides in individuals, based upon their stock of knowledge skills, health and nutrition. It is enhanced by access to services that provide these, such as schools, medical services and adult training. People's productivity is increased by their capacity to interact with productive technologies and with other people. Leadership and organizational skills are particularly important in making other resources more valuable. *Physical capital* is the store of human-made material resources, and comprises buildings, such as housing and factories, market infrastructure, irrigation works, roads and bridges, tools and tractors, communications, and energy and transportation systems. All of these resources make labour more productive. *Financial capital* is more of an accounting concept: it serves as a facilitating role, rather than as a source of productivity in and of itself. It represents accumulated claims on goods and services, built up through financial systems that gather savings and issue credit, such as pensions, remittances, welfare payments, grants and subsidies.

As agricultural systems shape the very assets upon which they rely for inputs, a vital feedback loop occurs from outcomes to inputs. Donald Worster's three principles for good farming capture this idea. Good farming makes people healthier, promotes a more just society, and preserves the Earth and its networks of life. He says: *'the need for a new agriculture does not absolve us from the moral duty and common-sense advice to farm in an ecologically rational way. Good farming protects the land, even when it uses it'.*[5] Thus, sustainable agricultural systems tend to have a positive effect on natural, social and human capital, while unsustainable ones feed back to deplete these assets, leaving less for future generations. For example, an agricultural system that erodes soil while producing food externalizes costs that others must bear. But a system that sequesters carbon in soils through organic matter accumulation helps to mediate climate change. Similarly, a diverse agricultural system that enhances on-farm wildlife for pest control contributes to wider stocks of biodiversity, while simplified modernized systems that eliminate wildlife do not. Agricultural systems that offer labour-absorption opportunities, through resource improvements or value-added activities, can boost economies and help to reverse rural-to-urban migration patterns.

Agriculture is, therefore, fundamentally multifunctional. It jointly produces many unique non-food functions that cannot be produced by other economic sectors as efficiently. Clearly, a key policy challenge, for both industrialized and developing countries, is to find ways in which to maintain and enhance food production. But the key question is: can this be done while improving the positive side effects and eliminating the negative ones? It will not be easy, as past agricultural development has tended to ignore both the multifunctionality of agriculture and the pervasive external costs.[6]

This leads us to a simple and clear definition of sustainable agriculture. It is farming that makes the best use of nature's goods and services while not damaging the environment.[7] Sustainable farming does this by integrating natural processes, such as nutrient cycling, nitrogen fixation, soil regeneration and natural pest control, within food production processes. It also minimizes the use of non-renewable inputs that damage the environment or harm the health of farmers and consumers. It makes better use of farmers' knowledge and skills, thereby improving their self-reliance, and it makes productive use of people's capacities to work together in order to solve common management problems. Through this, sustainable agriculture also contributes to a range of *public goods*, such as clean water, wildlife, carbon sequestration in soils, flood protection and landscape quality.

## Putting Monetary Values on Externalities

Most economic activities affect the environment, either through the use of natural resources as an input or by using the 'clean' environment as a sink for pollution. The costs of using the environment in this way are called *externalities*. Because externalities comprise the side effects of economic activity, they are external to markets, and so their costs are not part of the prices paid by producers or consumers. When such externalities are not included in prices, they distort the market by encouraging activities that are costly to society, even if the private benefits are substantial. The types of externalities encountered in the agricultural sector have several features. Their costs are often neglected, and often occur with a time lag. They often damage groups whose interests are not represented, and the identity of the producer of the *externality* is not always known.[8]

In practice, there is little agreed data on the economic cost of agricultural externalities. This is partly because the costs are highly dispersed and affect many sectors of economies. It is also necessary to know about the value of nature's goods and services, and what happens when these largely unmarketed goods are lost. Since the current system of economic accounting grossly underestimates the current and future value of natural capital, this makes the task even more difficult.[9] It is relatively easy, for example, to count the remedial treatment costs that follow pollution incidents; but it is much more difficult to value, for example, skylarks singing on a summer's day, and the costs incurred when they are lost.

Several studies have recently put a cost on the negative externalities of agriculture in China, Germany, the Netherlands, the Philippines, the UK and the US.[10] When it is possible to make the calculations, our understanding of what is the best or most efficient form of agriculture can change rapidly. In the Philippines, researchers from the International Rice Research Institute found that modern rice cultivation was costly to human health. They investigated the health status of rice farmers who were exposed to pesticides, and estimated the monetary costs of significantly increased incidence of eye, skin, lung and neurological disorders. By incorporating these within the economics of pest control, they found that modern high-input pesticide systems suffer twice. For example, with nine pesticide sprays per season, they returned less per *hectare* than the *integrated pest management* strategies and cost the most in terms of ill health. Any expected positive production benefits of applying pesticides were overwhelmed by the health costs. Rice production using natural control methods exhibits *multifunctionality* by contributing positively both to human health and by sustaining food production.[11]

At the University of Essex, we recently developed a new framework to study the negative externalities of UK agriculture. This framework uses seven cost categories to assess negative environmental and health costs, such as damage to water, air, soil and biodiversity, and damage to human health by pesticides, micro-organisms and disease agents. The analysis of damage and monitoring costs counted only external costs; private costs borne by farmers themselves, such as increased pest or weed resistance from pesticide overuse, were not included. We conservatively estimated that the external costs of UK agriculture, almost all of which is modernized and industrialized, were at least UK£1.5 billion to UK£2 billion each year. Another study by Olivia Hartridge and David Pearce has also put the annual costs of modern agriculture in excess of UK£1 billion.[12] These are costs imposed on the rest of society and are, effectively, a hidden subsidy to the polluters.[13] The annual costs arise from damage to the atmosphere (UK£316 million), to water (UK£231 million), to bio-diversity and landscapes (UK£126 million), to soils (UK£96 million), and to human health (UK£777 million). Using a similar framework of analysis, the external costs in the US amount to nearly UK£13 billion per year.[14]

How do all of these costs arise? Pesticides, nitrogen and phosphorus nutrients, soil, farm wastes and micro-organisms escape from farms to pollute ground and surface water. Costs are incurred by water delivery companies, and then passed on to their customers in order to remove these contaminants, to pay for restoring watercourses following pollution incidents and *eutrophication*, and to remove soil from water. Using UK water companies' returns for both capital and operating expenditure, we estimated annual external costs to be UK£125 million for the removal of pesticides below legal standards, UK£16 million for nitrates, UK£69 million for soil, and UK£23 million for *Cryptosporidium*.[15] These costs would be much greater if the policy goal were complete removal of all contamination.

Agriculture also contributes to atmospheric pollution through the emissions of four gases: methane from livestock, nitrous oxide from fertilizers, ammonia from livestock wastes and some fertilizers, and carbon dioxide from energy and fossil-fuel consumption and the loss of soil carbon. These, in turn, contribute to atmospheric warming (methane, nitrous oxide and carbon dioxide), ozone loss in the stratosphere (nitrous oxide), acidification of soils and water (ammonia) and eutrophication (ammonia). The annual cost for these gases is some UK£444 million.[16] A healthy soil is vital for agriculture; but modern farming has accelerated erosion, primarily through the cultivation of winter cereals, the conversion of pasture to arable, the removal of field boundaries and hedgerows, and overstocking of livestock on grasslands. Off-site costs arise when soil washed or blown away from fields blocks ditches and roads, damages property, induces traffic accidents, increases the risk of floods, and pollutes water through sediments and associated nitrates, phosphates and pesticides. These costs amount to UK£14 million per year. Carbon in organic matter in soils is also rapidly lost when pastures are ploughed or when agricultural land is intensively cultivated, and adds another UK£82 million to the annual external costs.

Modern farming has had a severe impact on wildlife in the UK. More than nine-tenths of wildflower-rich meadows have been lost since the

1940s, together with one half of heathland, lowland *fens*, and valley and basin mires, and one third to one half of ancient lowland woods and hedgerows. Species diversity is also declining in the farmed habitat itself. Increased use of drainage and fertilizers has led to grass monocultures replacing flower-rich meadows; overgrazing of uplands has reduced species diversity; and herbicides have cut diversity in arable fields. Hedgerows were removed at a rate of 18,000 kilometres a year between the 1980s and 1990s. Farmland birds have particularly suffered, with the populations of nine species falling by more than one half in the 25 years to 1995.[17] The costs of restoring species and habitats under biodiversity action plans were used as a proxy for the costs of wildlife and habitat losses; together with the costs of replacing hedgerows, stonewalls and bee colonies, this brings the annual costs to UK£126 million.

Pesticides can affect workers who are engaged in their manufacture, transport and disposal, operators who apply them in the field, and the general public. But there is still great uncertainty because of differing risks per product, poor understanding of chronic effects (such as in cancer causation), weak monitoring systems, and misdiagnoses by doctors.[18] For these reasons, it is very difficult to say exactly how many people are affected by pesticides each year. According to voluntary reporting to government, 100–200 incidents occur each year in the UK.[19] However, a recent government survey of 2000 pesticide users found that 5 per cent reported at least one symptom in the past year about which they had consulted a doctor, and a further 10 per cent had been affected, mostly by headaches, but had not consulted a doctor, incurring annual costs of about UK£1 million. Chronic health hazards associated with pesticides are even more difficult to assess. Pesticides are ingested via food and water, and these represent some risk to the public. With current scientific knowledge, it is impossible to state categorically whether or not certain pesticides play a role in cancer causation. Other serious health problems arising from agriculture are food-borne illnesses, antibiotic resistance and BSE-CJD.[20]

These external costs of UK agriculture are alarming. They should call into question what we mean by efficiency. Farming receives UK£3 billion of public subsidies each year, yet causes another UK£1.5 billion of costs elsewhere in the economy. If we had no alternatives, then we would have to accept these costs. But in every case, there are choices. Pesticides do not have to get into watercourses. Indeed, they do not need to be used at all in many farm systems. The pesticide market in the UK is UK£500 million; yet, we pay UK£120 million just to clean them out of drinking water. We do not need farming that damages biodiversity and landscapes; we do not need intensive livestock production that encourages infections and overuse of antibiotics. Not all costs, though, are subject to immediate elimination with sustainable methods of production. Cows will still belch methane, until animal feed scientists find a way of amending ruminant biochemistry to prevent its emission. But it is clear that many of these massive distortions could be removed with some clear thinking, firm policies, and brave action by farmers.[21]

## The Side Effects of Intensive Food Production on Water and Wetlands

One problem with the redesign of landscape for modern agriculture is that important natural features and functions are lost. Watercourses are one of the most tamed and abused of natural landscape features. Wetlands have been drained, rivers straightened or hidden behind levees, aquifers mined, and rivers, lakes and seas polluted, mostly to ensure that productive farmland is protected from harm or excessive costs. Once again, the narrow view that farmland is only important for food production has caused secondary problems. According to the National Research Council, 47 million hectares of wetlands in the US were drained during the past two centuries, and 85 per cent of inland waters are now artificially controlled. This created new farmland, to the benefit of farmers. But remove the wetlands, and the many valued services they provide are also lost. They are habitats for biodiversity, capture nutrients that run off fields, provide flood protection, and are important cultural features of the landscape.

Donald Worster describes growing up within 30 metres of the already tamed Cow Creek in Kansas: *'We could not see it from our windows; we could only see the levee.'* During the 19th century, the town expanded by the river and the early settlers converted land to wheat cultivation. As a result, the natural and regular flooding of the river started to cause considerable economic damage. Episodes of flooding and continued expenditure on flood protection continued for decades, until a major flood in 1941 finally led the Army Corps of Engineers to construct a series of 4-metre high levees: *'Now at long last the good Kansas folk, having vanquished the Indians and the bison and the sandhill cranes and the antelope, had managed to vanquish Cow Creek. Abruptly, it disappeared from their lives.'* This is the alarming part. When valued landscape features have gone, or have been replaced, the everyday experiences of local people will steadily erode old memories. The young will not know, while the old will be troubled, until they pass on, too.[22] Meanwhile, we all lose.

In Europe, river valleys used to contain many water meadows. These fields were likely to be flooded by overflowing rivers, and were used productively in order to produce a late winter or early spring crop of grass. More importantly, when the river did flood, water was stored on the meadows and did not harm housing or other vulnerable areas. However, in the intensification of food production, most of these meadows were converted to arable fields. At the same time, rivers were tamed through channelling, field sizes increased, hedgerows removed, and houses built on vulnerable land. Now, when it rains, the consequence is increased flooding to vulnerable areas. It looks as though there has been 'too much' rain; but, in truth, this is largely due to changes in the landscape.

In Germany, Rienk van der Ploeg and colleagues have correlated loss of meadows with an increased incidence of inland floods. Over a century, 6 of the 12 most extreme events have occurred since 1983. They show that changes in the diversity of the use of agricultural land are the main cause of flooding. In particular, permanent meadows have been converted to arable fields, some 1.5 million hectares since the mid 1960s. Surface sealing and compaction means these fields are less likely to hold water during winter. Another 4.5 million hectares of wetland soils have been drained since the 1940s. Thus, when it rains, water contributes more rapidly to river water discharge, thereby increasing the likelihood of flooding. The cost of two floods in 1993 and 1995 was nearly 2 billion Deutschmarks, and van der Ploeg concludes that the conversion of arable back to permanent meadows would be economically and environmentally beneficial: *'It must be acknowledged that any further increase in agricultural productivity is likely to cause additional adverse environmental effects. Future farm policy must pay more attention to the environment'.*[23]

Japan provides another example of the wider value of agricultural wetlands – in this case, irrigated paddy rice fields. Japan's very high rainfall is concentrated into a few months within a landscape characterized by a high mountain chain. With a very short flow time to the sea, this means that much of the country is subject to severe flood risk. Paddy rice farming, though, provides an important sink for this water. There are more than 2 million hectares of paddy rice in Japan, and each of these hectares holds about 1000 tonnes of water each year. In the Koshigaya City basin, 25 kilometres north of Tokyo, paddy fields close to the city have been steadily converted to residential uses over the past quarter century. But as the area of paddy has declined by about 1000 hectares since the mid 1970s, so the incidence of flooding has increased. Each year, 1000 to 3000 houses are flooded. In whole *watersheds*, woods and farms on steep slopes have been identified as having the greatest value in buffering and slowing water flow, and minimizing landslides. Diversity, though, is critical. As Yoshitake Kato and colleagues have put it:

*Traditional villages in rural areas include settlements, paddy fields, crop fields and forested hills or mountains, all as linked landscape. The systems were dependent on all their parts. The decline of farming in the uplands, together with loss of forests, threatens the stability of whole watersheds.*[24]

In China, the 500,000 hectares of wetlands that have been reclaimed for crop production during the past 50 years have meant the loss of flood-water storage capacity of some 50 billion cubic metres, a major reason for the US$20 billion flood damage caused in 1998.[25] In many agricultural systems, over-intensive use of the land has resulted in sharp declines in soil organic matter and/or increases in soil erosion, some of which, in turn, threatens the viability of agriculture itself. In South Asia, for example, one quarter of farmland is affected by water erosion, one fifth by wind erosion, and one sixth by *salinization* and waterlogging.[26]

Putting a value on wetlands and watercourses, so that we can calculate how much is lost when they are damaged or destroyed, is not a trivial task. Economists have no agreed value for wetlands, though various studies indicate that individual bodies can provide several million dollars of free services to nearby communities for waste assimilation and treatment. A recent US Department of Agriculture study put wetland monetary value at US$300,000 per hectare per year. Another way to assess value is to investigate how much people pay to visit wetlands, whether to watch or photograph biodiversity, or indeed to shoot it. In the US, it is estimated that 50 million people each year spend US$10 billion observing and photographing wetland flora and fauna, 31 million anglers spend US$16 billion on fishing, and 3 million waterfowl hunters spend nearly US$700 million dollars annually on shooting it. A recent meta-analysis of economic studies of people's willingness to pay for recreational services of wetlands and watercourses puts the average value in Europe at UK£20 to UK£25 per person per hectare per year.[27] Thus, each hectare of wetland converted to another purpose means the loss of at least UK£20 of value to the public. There are, of course, limitations in these exercises, as monetary values cannot be allocated to all uses.

One of the most serious side effects of agriculture is the leaching and run-off of nutrients, and their disruption of water ecosystems. *Eutrophication* is the term used to describe nutrient enrichment of water that leads to excessive algal growth, disruption of whole food webs and, in the worst cases, complete eradication of all life through deoxygenation. The most notorious example is the Gulf of Mexico dead zone, an area of 5000 to 18,000 square kilometres of sea that has received so much nutrient input that all aquatic life has been killed. The cost of farm overuse of nutrients in the Mississippi Basin is thus borne by the fishing families of Louisiana.

No one has yet put a cost on these losses. However, if they were internalized in the prices of fertilizers, or the activities of intensive livestock units, we would expect much greater concern about such polluting activities.[28]

At the University of Essex, we recently conducted a study of the costs of nutrient enrichment of water in the UK.[29] Eutrophication affects the value of waterside properties, and reduces the recreational and amenity value of water bodies for water sports, angling and general amenity; for industrial uses; for the tourist industry; and for commercial aquaculture, fisheries, and shell-fisheries. Additional costs are incurred through a variety of social responses by both statutory and non-statutory agencies. In total, we estimate nutrient enrichment to cost some UK£130–170 million per year in the UK.[30]

## Industrialized Agriculture and Food-Borne Illnesses

Having mostly conquered hunger in industrialized countries, it is a sad irony that food is now a major source of ill health. We eat too much, we eat the wrong mix of foods, and we get ill from food-borne illnesses. In Europe, 10 to 20 per cent of all people are defined as obese, with a body mass index greater than 30 kilogrammes per square metre. The World Health Organization (WHO) estimates that 2 to 7 per cent of healthcare costs in Europe arise from obesity, and one American study suggests that a 10 per cent weight loss amongst obese people would increase life expectancy by two to seven months, and produce lifetime benefits of US$2000 to US$6000 per person.[31] Several diseases are strongly linked to unbalanced food consumption, including non-insulin dependent diabetes, the incidence of which is growing rapidly, together with strokes, coronary heart disease and some cancers.

Many of these health problems, though, are attributable to the choices consumers make. We could eat five portions of fruit and vegetables per day, thus protecting against many of these problems, but for a variety of reasons we do not. But we cannot choose when it comes to food-borne diseases. The WHO estimates that 130 million people in Europe are affected by food-borne diseases each year, mainly from biological sources, particularly strains of *Salmonella*, *Campylobacter*, *Listeria* and *E.coli*. *Salmonella* is the most common pathogen, accounting for up to 90 per cent of cases in some countries. Throughout the world, diarrhoea is the most common symptom of food-borne illness, and is a major cause of death and retardation of growth in infants. There is evidence that cases of *Campylobacter* and *Salmonella* poisoning are increasing in Europe, though some of the increases can be explained by better monitoring systems.[32] In the US, the incidence of food-borne illness is greater, perhaps because of the greater industrialization of agriculture and, in particular, of livestock raising. According to the US government's Centres for Disease Control, 76 million people in the US fall ill each year from food-borne illness, of whom more than 300,000 are hospitalized and 5000 die.[33]

The costs of food-borne illnesses are massive. The Institute for Medicine at the National Academy of Sciences, the US Department of Agriculture and the World Health Organization estimate such illnesses in the US to cost between US$34 billion and US$110 billion per year. In the UK, the government's Food Standards Agency estimates that each of the annual 5 million cases of food poisoning costs on average UK£85, comprising costs to health services and losses to businesses, putting the annual cost at more than UK£400 million pounds. These data suggest that one in four Americans and one in ten Britons suffer from food poisoning each year.[34]

Some of these food-borne illnesses arise from shellfish, others are associated with mass catering or occur in the food processing chain. But it is the initial sources of infection on the farm, combined with the overuse of antibiotics for growth promotion, that is an increasing source of disquiet. The concentration of livestock into factory feedlots, broiler sheds and colossal pig units promotes infection and spread. As the WHO puts it: *'The greatest risk appears to be the production of animal foods. It is from this source that the most serious health threats originate, for instance, Salmonella, Campylobacter, E.coli and Yersinia.'* The pool of infection at the start of the food chain is now very serious. The US Department of Agriculture has found very high levels of microbial infections in US farm animals, particularly in broiler chickens and turkeys. *Clostridium* has been found in 30 to 40 per cent of flocks, *Campylobacter* in nearly 90 per cent, *Salmonella* in 20 to 30 per cent, and *Staphylococcus* in 65 per cent. These levels of infection are matched in some European countries, with more than 90 per cent of pig herds and nearly 50 per cent of cattle in the Netherlands and Denmark contaminated with *Campylobacter*. At these levels of incidence in animals, it is hardly surprising that illnesses from meat consumption are so common. Incidences of illness in pigs and cattle in the US are much lower, but still a worrying 3 to 30 per cent of herds for these four pathogens.[35]

This extraordinary problem, which underlies the desire for ever-cheaper foods, is worsened by antibiotic resistance, brought on by overuse of antibiotics for livestock growth promotion and over-prescription in medicine. Twenty-three thousand tonnes of antibiotics are used in the US each year, of which 11,000 are given to animals, four-fifths of which is just for growth promotion. In the UK, 1200 tonnes of antibiotics are used each year, 40 per cent of which is for humans, 30 per cent for farm

animals, and 30 per cent for domestic pets and horses. Only one fifth of
the antibiotics and other *antimicrobials* that are used in modern agriculture
are for therapeutic treatment of clinical diseases, with four-fifths for
prophylactic use and growth promotion. The US Centres for Disease
Control say: *'antimicrobial resistance is a serious clinical and public health problem in
the US'*, and one estimate from the Institute of Medicine suggests that such
resistance costs US$30 million per year. A UK House of Lords select
committee enquiry was even more alarmed, recently stating: *'There is a
continuing threat to human health from the imprudent use of antibiotics in animals. . .
we may face the dire prospect of revisiting the pre-antibiotic era.*[86]

In both Europe and North America, the most common forms of
antimicrobial resistance are to strains of antibiotics used in treating
animals, and these are transferred to human patients. Some antibiotics,
such as fluoroquinones and avoparcin (used to treat infections in poultry
and as growth promoters), are now associated with dramatic increases in
resistant diseases in humans. Fluoroquinone resistance is thought to be
the main factor why *Campylobacter* infections have become so common in
the Netherlands. As the WHO puts it: *'Campylobacter species are now the
commonest cause of bacterial gasteroenteritis in developed countries, and cases are
predominantly associated with consumption of poultry.'*[37] There is no such thing as
a cheap chicken.

## Putting a Monetary Value on Agricultural Landscapes

Landscapes are culturally valuable, and the aesthetic value we gain from
them owes much to their emergence from agricultural practices. They are,
of course, almost impossible to value in monetary terms. However, many
proxies can be used, including how much governments are willing to pay
farmers to produce certain habitats or landscapes, how often the public
visits the countryside, and how much they spend when they get there. In
the UK, several studies of agri-environmental policies have sought to put
a value on positive environmental and landscape outcomes.[38] These
schemes have attempted to restore some of the habitats and other positive
countryside attributes that were lost during intensification, as well as to
protect those attributes not yet lost.

UK agri-environmental schemes have been designed to deliver benefits
in several forms, including biodiversity, landscape patterns, water quality,
archaeological sites, and enhanced access. Benefits may accrue to those in
the immediate area of a scheme, to visitors from outside the area, and to

the public at large. The annual per-household benefits, using a variety of valuation methods such as contingent valuation, choice experiments and contingent ranking, vary from UK£2 to UK£30 for most environmentally sensitive areas (ESAs), rising to UK£140 for the Norfolk Broads and UK£380 for Scottish machair grasslands. If we take the range of annual benefits per household to be UK£10 to UK£30, and assume that this is representative of the average household's preferences for all landscapes produced by agriculture, then this suggests national benefits of the order of UK£200 million to UK£600 million. Expressed on a per hectare basis, this suggests annual benefits of UK£20 to UK£60 per hectare of arable and pasture land in the UK.

On the one hand, these are likely to be overestimates, assuming agri-environment schemes have already targeted certain landscapes because of their higher value. On the other hand, they could be substantial under-estimates, as they do not value such benefits as pathogen-free foods, uneroded soils, emission-free agriculture, and biodiversity-producing systems. They also focus on the outcomes of a scheme rather than on the whole landscape. There are too few studies to corroborate these data. One study in the UK compared paired organic and non-organic farms, and concluded that organic agriculture produces UK£75 to UK£125 per hectare of positive externalities each year, with particular benefits for soil health and wildlife.[39] As there are 3 million hectares of organic farming in Europe, the annual positive externalities could be UK£300 million, assuming that benefits hold for the many organic farming systems across Europe.

Actual visits made to the countryside are another proxy measure of how much we value landscapes. Each year in the UK, day and overnight visitors make some 433 million visit-days to the countryside and another 118 million to the seaside.[40] The average spend per day or night varies from nearly UK£17 for UK day visitors, to UK£33 for UK overnight visitors, and just over UK£58 for overseas overnight visitors. This indicates that the 551 million visit-days to the countryside and seaside result in spending of UK£14 billion per year. This is 3.5 times greater than the annual public subsidy of farming, and indicates just how much we value the landscape.

If it is clean water that is required, the value of an agricultural landscape can be substantial – as New York State has found out with its support for sustainable agriculture in the 500,000 hectare Catskill-Delaware watershed complex.[41] New York City gets 90 per cent of its drinking water from these watersheds, some 6 billion litres a day. In the late 1980s, though, the city was faced with having to construct a filtration facility in order to meet new drinking water standards, the cost of which would be US$5 billion to US$8 billion dollars, plus another US$200 million to

US$500 million dollars in annual operating costs. One third of the cropland in the watershed would have to be taken out of farming in order to reduce run-off of eroded soil, pesticides, nutrients and bacterial and protozoan pathogens.

Instead, the city opted for a collaborative approach with farmers. It supported the establishment of a Watershed Agricultural Council in the early 1990s, a partnership between farmers, government and private organizations with the dual aim of protecting the city's drinking water supply and sustaining the rural economy. It works on whole-farm planning with each farm, tailoring solutions to local conditions in order to maximize reductions in off-site costs. The first two phases of the programme, leading to the 85 per cent target in pollution reduction, cost some US$100 million, a small proportion of the cost of the filtration plant and its annual costs. Not only do taxpayers benefit from this approach to joint agri-environmental management, but so do farmers, the environment and rural economies.[42] The only surprising thing is that these initiatives are still rare.

## Agriculture's Carbon Dividend

The greatest environmental problem we face anywhere in the world now is climate change provoked by rising levels of anthropogenic greenhouse gases. Climate change threatens to disrupt economies and ecosystems, to challenge existing land uses, to substantially raise sea levels, and to drown coastal lands and even some whole countries. In order to slow down and eventually to reverse these changes, we need to reduce human-induced emissions of these gases, as well as to find ways of capturing or locking up carbon from the atmosphere. Sustainable agriculture can make an important contribution to climate change mitigation through both emissions reductions and carbon sequestration. As the international markets for carbon expand, so sequestered carbon could represent an important new income source for farmers.[43]

Agricultural systems contribute to carbon emissions through the direct use of fossil fuels in farm operations, the indirect use of embodied energy in inputs that are energy intensive to manufacture and transport (particularly fertilizers and pesticides), and the cultivation of soils resulting in the loss of soil organic matter. Agriculture is also an accumulator of carbon, offsetting losses when organic matter is accumulated in the soil, or when above-ground woody biomass acts either as a permanent sink or is used as an energy source that substitutes for fossil fuels.

Long-term agricultural experiments in both Europe and North America indicate that soil organic matter and soil carbon are lost during intensive cultivation. But both can be increased with sustainable management practices. The greatest dividend comes from the conversion of arable to agroforestry: there is a benefit from both increased soil organic matter and the accumulation of above-ground woody biomass. Grasslands within rotations, *zero-tillage* farming, the use of legumes and green manures, and high amendments of straw and manures, also lead to substantial carbon sequestration. There is now good evidence to show that sustainable agricultural systems can lead to the annual accumulation of 300 to 600 kilogrammes of carbon per hectare, rising to several tonnes per hectare when trees are intercropped in cropping and grazing systems.

Agriculture as an economic sector also contributes to carbon emissions through the consumption of direct and indirect fossil fuels. With the increased use of nitrogen fertilizers, pumped irrigation and mechanical power, accounting for more than 90 per cent of the total energy inputs to farming, industrialized agriculture has become progressively less energy efficient. The difference between sustainable and conventional systems of production is striking. Low-input or organic rice in Bangladesh, China, and Latin America is some 15 to 25 times more energy efficient than irrigated rice grown in the US. For each tonne of cereal or vegetable from industrialized high-input systems in Europe, 3000 to 10,000 *megajoules* of energy are consumed in its production. But for each tonne of cereal or vegetable from sustainable farming, only 500 to 1000 megajoules are consumed.[44]

It is now known that intensive cultivation of cereals leads to reductions in soil organic matter and carbon content. However, recent years have seen an extraordinary growth in the adoption of conservation tillage and zero-tillage systems, particularly in the Americas. These systems of cultivation maintain a permanent or semi-permanent organic cover on the soil. The function is to protect the soil physically from the action of sun, rain and wind, and to feed soil biota. The result is reduced soil erosion and improved soil organic matter and carbon content. Zero-tillage systems and those using legumes as green manures and/or cover crops contribute to organic matter and carbon accumulation in the soil. Zero-till systems also have an additional benefit of requiring less fossil fuel for machinery passes. Intensive arable with zero-tillage results in the annual accumulation of 300 to 600 kilogrammes of carbon per hectare. With mixed rotations and cover crops, this system can accumulate up to 1300 kilogrammes of carbon per hectare.

The 1997 Kyoto Protocol to the UN Framework Convention on Climate Change established an international policy context for the reduction of carbon emissions and the expansion of carbon sinks in order to address climate change. Under the protocol and the 2001 Bonn and Marakesh agreement, the principle of financial and technological transfers to land management projects and initiatives was established. Article 17 permits countries to produce certified emissions reductions (also known as offsets) and emissions reductions units through joint implementation projects. Since it is cheaper for many countries to abate greenhouse gas emissions, working together for joint implementation is, in theory, a cost-effective mechanism for achieving global targets.

Nevertheless, for real impacts on climate change to occur, sinks must become permanent. If lands under conservation tillage are ploughed, then all the gains in soil carbon and organic matter are lost. This poses a big challenge for trading systems, as there is no such thing as a permanent emissions reduction or a permanently sequestered tonne of carbon. Despite these uncertainties, carbon banks, boards of trade, and trading systems emerged during the year 2000. The early carbon trading systems set per-tonne credit values mostly in the US$2 to US$10 range, though the real value of each tonne sequestered is much higher. The important policy questions centre on how to establish permanent or indefinite sinks, how to prevent leakage, such as re-ploughing of zero-tilled fields or deforestation, how to agree measurements, and whether the cost of implementation can be justified as a result of additional side effects or *multifunctionality*.

We do not yet know how much carbon could be locked up in response to monetary incentives for carbon sequestration. The empirical evidence is relatively sparse, and practical experience even more limited. No agreed system of payment levels has yet been established. Another unresolved issue relates to the location for the greatest carbon returns on investments. Investments in creating sustainable systems in the tropics are likely to be cheaper than in temperate regions, where industrialized agriculture prevails. Such financial transfers from industrialized to developing countries could produce substantial net global benefits, as well as benefit poor farmers. At current prices, it is clear that farmers will not solely become 'carbon farmers'. However, systems that accumulate carbon are also delivering many other public goods, such as improved biodiversity and clean water from watersheds. Policy-makers may also seek to price these in order to increase the total payment package. Carbon, therefore, represents an important new source of income for farmers, as well as encouraging them to adopt sustainable practices.

# Could Better Policies Help?

These external costs and benefits of agriculture raise important policy questions. In particular, should farmers receive public support for the public benefits they produce in addition to food? Should those individuals and organizations who pollute have to pay for restoring the environment and human health? These two principles are called 'the provider gets' and 'the polluter pays', and they are important to both industrialized and developing countries. Three categories of policy instruments are available: advisory and institutional measures; regulatory and legal measures; and economic instruments. In practice, effective pollution control and the supply of desired public goods requires a mix of all three approaches, together with integration across sectors.

Advisory and institutional measures have long formed the backbone of policies to internalize costs and so prevent agricultural pollution. These measures rely on the voluntary actions of farmers, and are favoured by policy-makers because they are cheap and adaptable. Advice is commonly given in the form of codes of good agricultural practice, such as recommended rates of pesticide and fertilizer application, or measures for soil erosion control. Most governments still employ extension agents who work with farmers on technology development and transfer. A variety of institutional mechanisms can also help to increase social capital and the uptake of more sustainable practices, including encouraging farmers to work together in study groups, investing in extension and advisory services to encourage greater interaction between farmers and *extensionists*, and encouraging new partnerships between farmers and other rural *stakeholders*, since regular exchanges and reciprocity increase trust and confidence, and lubricate cooperation.

Regulatory and legal measures are also used to internalize external costs. This can be done either by setting emissions standards for the discharge of a pollutant, or by establishing quality standards for the environment receiving the pollutant. Polluters who exceed standards are then subject to penalties. There are many types of standards, such as operating standards to protect workers; production standards to limit levels of contaminants of residues in foods; emissions standards to limit releases or discharges, such as silage effluents; and environmental quality standards for undesirable pollutants in vulnerable environments, such as pesticides in water. But the problem with such regulations is that most agricultural pollutants are diffuse, or non-point, in nature. It is impossible for inspectors to ensure compliance on hundreds of thousands of farms in the way that they can with a small number of factories. Regulations are also used to eliminate certain practices, and include bans on the spraying of pesticides close to rivers and on straw-burning in the UK, as well as the mandatory requirement to complete full nutrient accounts for farms, such as in the Netherlands and Switzerland. A final use for regulations is the designation and legal protection of certain habitats and species, which are set at national or international levels.

Economic instruments can be used to ensure that the polluter bears the costs of the pollution damage and the abatement costs incurred in controlling the pollution. They can also be used to reward good behaviour. A variety of economic instruments are available for achieving internalization, including environmental taxes and charges, tradable permits, and the targeted use of public subsidies and incentives. Environmental taxes seek to shift the burden of taxation away from economic goods, such as labour, and towards environmental bads, such as waste and pollution. Clearly, the market prices for agricultural inputs do not currently reflect the full costs of their use. Environmental taxes or pollution payments, however, seek to internalize some of these costs, in this way encouraging individuals and businesses to use them more efficiently. Such taxes offer the opportunity of a double dividend by cutting environmental damage, particularly from non-point sources of pollution, while promoting welfare. However, many opponents still believe that environmental taxes stifle economic growth, despite compelling evidence to the contrary.[45]

There is now a wide range of environmental taxes used by countries of Europe and North America. These include carbon and energy taxes in Belgium, Denmark and Sweden; chlorofluorocarbon taxes in Denmark and the US; sulphur taxes in Denmark, France, Finland and Sweden; nitrogen oxide charges in France and Sweden; leaded and unleaded petrol differentials in all European Union countries; landfill taxes in Denmark, the Netherlands and the UK; groundwater extraction charges in the Netherlands; and sewage charges in Spain and Sweden. However, environmental taxes have rarely been applied to agriculture, with the notable exception of pesticide taxes in Denmark, Finland, Sweden and in several US states; fertilizer taxes in Austria, Finland, Sweden, and several US states; and manure charges in Belgium and the Netherlands.[46]

The alternative to penalizing farmers through taxation is to encourage them to adopt non-polluting technologies and practices. This can be done by offering direct subsidies for the adoption of sustainable technologies, and by removing perverse subsidies that currently encourage polluting activities.[47] An important policy principle suggests that it is more efficient to promote practices that do not damage the environment, rather than spend money on cleaning up after a problem has been created. Many governments provide some direct or indirect public support to their domestic agricultural and rural sectors. Increasingly, payments are being shifted away from being production linked, such as through price support or direct payments, to being re-targeted to support sustainable practices. Generally, though, only small amounts of total budgets have been put aside for environmental improvements though such policies as the US Conservation Reserve Programme, the European Union's agri-environmental and rural development programmes, and the Australian Landcare programme. Many now believe that all public support for farming should be entirely linked to the provision of public environmental and social goods.

## The Radical Challenge of Integration

The substantial external costs of modern agriculture, and the known external benefits of sustainable agricultural systems, pose great challenges for policy-makers. A range of policy reforms could do much to internalize some of these costs and benefits in prices. In practice, since no single solution is likely to suffice, the key issue rests on how policy-makers choose an appropriate mix of solutions, how these are integrated, and how farmers, consumers and other stakeholders are involved in the process of reform itself. Attention will therefore need to be paid to the social and institutional processes that encourage farmers to work and learn together, and result in integrated cross-sectoral partnerships. Policy integration is vital; yet most policies seeking to link agriculture with more environmentally sensitive management are still highly fragmented.

The problem is that environmental policies have tended only to 'green' the edges of farming. Non-crop habitats have been improved, as have some hedgerows, woodlands and wetlands. But the food in modern farming is still largely produced in the conventional manner. The challenge is to find ways of substantially greening the middle of farming – in the field rather than around the edges. A thriving and sustainable agricultural sector requires both integrated action by farmers and communities, and integrated action by policy-makers and planners. This implies horizontal integration with better linkages between sectors, and vertical integration with better linkages from the micro to macro level. Most policy initiatives are still piecemeal, affecting only a small part of an individual farmer's practices, and therefore not necessarily leading to substantial shifts towards sustainability.

The 1990s saw considerable global progress towards recognizing the need for policies to support sustainable agriculture. In a few countries, this has been translated into supportive and integrated policy frameworks.

In most, however, sustainability policies remain at the margins. Only two countries have given explicit national support for sustainable agriculture, putting it at the centre of agricultural development policy. Several countries have given sub-regional support, such as the states of Santa Caterina, Paraná and Rio Grande do Sol in southern Brazil who support zero-tillage and catchment management, and some states in India who support watershed management or participatory irrigation management. Many more countries have reformed parts of agricultural policies, such as China's support for integrated ecological demonstration villages; Kenya's catchment approach to soil conservation; Indonesia's ban on pesticides and its programme for farmer field schools; India's support for soybean processing and marketing; Bolivia's regional integration of agricultural and rural policies; Sweden's support for organic agriculture; Burkina Faso's land policy; and Sri Lanka's and the Philippines' stipulation that water users' groups manage irrigation systems.

One of the best examples of a carefully designed and integrated programme comes from China. In March 1994, the government published a White Paper to set out its plan for implementing Agenda 21. The plan advocated ecological farming, known as *Shengtai Nongye* or agroecological engineering, as the approach to achieve sustainable agriculture. Pilot projects have been established in some 2000 townships and villages spread across 150 counties. Policy for these 'eco-counties' is organized through a cross-ministry partnership, which uses a variety of incentives to encourage the adoption of diverse production systems to replace monocultures. These incentives include subsidies and loans, technical assistance, tax exemptions and deductions, security of land tenure, marketing services, and linkages to research organizations. These eco-counties contain some 12 million hectares of land, about half of which is cropland. Although this covers only a relatively small part of China's total agricultural land, it illustrates what is possible when policy is coordinated and holistic.

An even larger set of countries has seen some progress on agricultural sustainability at project and programme level. However, progress occurs in spite of, rather than because of, explicit policy support. No agriculture minister is likely to say that he or she is against sustainable agriculture; but wise words have yet to be translated into comprehensive policy reforms. Sustainable agricultural systems can be economically, environmentally and socially viable, and can contribute positively to local livelihoods. But without appropriate policy support, they are likely to remain, at best, localized in extent and, at worst, may simply wither away. In Europe and North America, most policy analysts and sustainable agriculture organizations now agree that a policy framework that integrates support for farming together with rural development and environmental protection could create new jobs, protect and improve natural resources, and support rural communities. Such a policy could include many of the elements seen in the progressive Swiss and Cuban policy reforms that were made during the 1990s.

## Cuba's National Policy for Sustainable Agriculture

At the turn of the century, Cuba was the only developing country with an explicit national policy for sustainable agriculture. Until the end of the 1980s, Cuba's agricultural sector was heavily subsidized by the Soviet bloc. Cuba imported more than half of all calories consumed, and 80 to 95 per cent of wheat, beans, fertilizers, pesticides and animal feed. It received three times the world price for its sugar. At the time, Cuba had the most scientists per head of population in Latin America, the most tractors per hectare, the second highest grain yields, the lowest infant mortality, the highest number of doctors per head of population, and the highest secondary school enrolment. But in 1990, trade with the Soviet bloc collapsed, leading to severe shortages in all imports, and restricting farmers' access to petroleum, fertilizers and pesticides.

The government's response was to declare an 'alternative model' as the official policy – an agriculture that focuses on technologies that substitute local knowledge, skills and resources for the imported inputs. It also emphasizes the diversification of agriculture, oxen to replace tractors, integrated pest management to replace pesticides, and the promotion of better cooperation among farmers, both within and between communities. The model has taken time to succeed. Calorific availability was 2600 kilocalories per day in 1990, fell disastrously to between 1000 and 1115 kilocalories per day soon after the transition, leading to severe hunger, but subsequently rose to 2700 kilocalories per day by the end of the 1990s.

Two important strands to sustainable agriculture in Cuba have emerged. Firstly, intensive organic gardens have been developed in urban areas – self-provisioning gardens in schools and workplaces (*autoconsumos*), raised container-bed gardens (*organoponicos*) and intensive community gardens (*huertos intensivos*). There are now more than 7000 urban gardens, and productivity has grown from 1.5 kilogrammes per square metre to nearly 20 kilogrammes per square metre. Secondly, sustainable agriculture is encouraged in rural areas, where the impact of the new policy has already been remarkable. More than 200 village-based and artisanal Centres for the Reproduction of Entomophages and Entomopathogens have been set up for biopesticide manufacture. Each year, they produce 1300 tonnes of *Bacillus thuringiensis* (*B.t.*) sprays for Lepidoptera control, nearly 800 tonnes of *Beaveria* sprays for beetle control, 200 tonnes of *Verticillium* for whitefly control, and 2800 tonnes of *Trichoderma*, a natural enemy of pests. Many biological control methods are proving more efficient than pesticides. Cut banana stems that are baited with honey to attract ants are placed in sweet potato fields, and have led to the control of sweet potato weevil. There are 170 vermi-compost centres, the annual production of which has grown from 3000 to 93,000 tonnes. Crop rotations, green manuring, intercropping and soil conservation have all been incorporated within polyculture farming.

At the forefront of the transition towards sustainable agriculture has been the Grupo de Agricultura Organica (GAO), formerly known as the Asociación Cubanes Agricultural Organica, which was formed in 1993. GAO brings together farmers, field managers, field experts, researchers and government officials to help spread the idea that organic-based alternatives can produce sufficient food for Cubans. Despite great progress, there remain many difficulties, including proving the success of the alternative system to sceptical farmers, scientists and policy-makers; developing new technologies sufficiently quickly to meet emergent problems; coordinating the many actors who work together; the need for continued decentralization of decision-making to farmer level; and appropriate land reform in order to encourage investment in natural asset-building.[48]

# The Swiss National Policy for Sustainable Agriculture

The Swiss Federal Agricultural Law was revised in 1992 in order to aim subsidies at ecological practices. It was then radically amended in 1996 following a national referendum in which 78 per cent of the public voted in favour of change.[49] The main priority was maintaining the important positive side effects of upland livestock farming – in particular, open meadows for skiing pistes in winter, but also the maintenance of rural mountain communities who are at the root of Swiss culture. Policy now differentiates between three different levels of public support. Tier one provides support for specific biotypes, such as extensive grasslands and meadows, high-stem fruit trees and hedges. Tier two supports integrated production with reduced inputs, meeting higher ecological standards than conventional farming. Tier three provides the most support for organic farming. As the directors of the federal agricultural and environmental offices, Hans Berger and Philippe Roch, have said: '*In ecological terms, Swiss agriculture is on the road to sustainability. There are encouraging signs that the agricultural reform has already began to have positive effects on nature and the environment.'*

Farmers must meet several minimum conditions in order to receive payments for integrated production, the so-called 'ecological standard' of performance. They must provide evidence that nutrient use matches crop demands, with livestock farmers having to sell surplus manures or reduce livestock numbers. Soils must be protected from erosion, and erosive crops, such as maize, can only be cultivated if alternated in rotation with meadows and green manures. At least 7 per cent of the farm must be allocated for species diversity protection through so-called 'ecological compensation areas', such as unfertilized meadows, hedgerows and orchards. Finally, pesticide use is restricted. A vital element of the policy process is that responsibility to set, administer and monitor is devolved to cantons, farmers' unions and farm advisors, local bodies and non-governmental organizations. By the end of the 1990s, 85 per cent of farmland complied with the basic ecological standard, which allows farmers to receive public subsidies. Some 5000 farms are now organic, and all farmers are soon expected to meet the ecological standard. Pesticide applications have fallen by one third in a decade, phosphate use is down by 60 per cent and nitrogen use by half. Semi-natural habitats have expanded during the decade, from 1 to 6 per cent in the plains, and from 7 to 23 per cent in the mountains.

There is much to learn from the Swiss and Cuban experiences, as these remain the only two countries at the turn of the century who put sustainable agriculture at the centre of their national policy. It is also true that Switzerland is a wealthy country and could afford to pay farmers for extra services. Cuba had no choice – it could not afford to do anything else. While it is difficult to draw general conclusions from these two cases, they highlight important questions. As American farmer and poet Wendell Berry pointed out: *'I cannot see why a healthful, dependable, ecologically sound farm- and farmer-conserving agricultural economy is not a primary goal of this country.'* Is there the political will in the remaining 200 or so countries for this kind of agriculture? The options are available, and the net benefits would be substantial. To date, the words have been easy, but the practice much more difficult.

## Concluding Comments

In this chapter, I have adopted a fairly narrow economic perspective in order to set out some of the real costs of modern agricultural and food systems. The side effects, or externalities, of food production systems are substantial; yet these do not appear in the price of food. The costs of lost biodiversity, water pollution, soil degradation, and ill health in humans are shifted elsewhere in economies. Because they are difficult to identify and measure, they are also easily lost. Allocating monetary values to these externalities is only one part of the picture because these methods are inevitably inexact. However, externalities do illustrate the size of the problem. The term multifunctional, when applied to agriculture, implies a system that does more than just produce food. Agriculture shapes landscapes, water quality, biodiversity and carbon stocks in soils. All of these are important *public goods*, and represent new income opportunities for farmers. But progress is slow, as policy reforms have lagged behind. There is a need for the radical integration of policies that support the transition towards agricultural systems which minimize their external costs and maximize their positive side effects.

## Chapter 4

# *Food for All*

## Revolutions in Central America

Elias Zelaya's hillside farm is found near a pine forest on the edge of the remote village of Pacayas in central Honduras. Fifteen years ago, the whole community was in the doldrums. Farms were poor-quality pasture and maize land, and many had been abandoned as worthless. No child in the village had ever been to secondary school. Land prices were low, and people saw their futures only in out-migration to the city. Yet, now local farmers are in the vanguard for diverse, sustainable and productive agriculture. In the mid 1980s, Elias happened to be in the right place and was lucky. He was encouraged to train as a farmer-*extensionist* by Roland Bunch and his colleagues at World Neighbors, and learned about low-cost, soil-improving technologies and how to adapt them to his own farm. The intercropping of legumes with maize immediately boosted cereal yields and improved soil health. Step by step, over the years, Elias added new enterprises to his farm, and there are now 28 types of crops and trees, together with pigs, chickens, rabbits, cattle and horses. Not all flourish – one day, an earthquake split the bottom of the fish pond. But most of the diverse enterprises are succeeding on this picturesque farm.

The effect is remarkable. The unimproved soils on the edge of Elias's farm are no more than a few centimetres deep, and beneath is hard bedrock. But in the fields where Elias grows legumes as green manures and uses composts, the soil is thick, dark, and spongy to the step. In some places on the farm, it is more than half a metre deep. No soils textbook will say this is possible, as soil is said to take thousands of years to create. Yet over a decade, Elias, and several tens of thousands of farmers in Central America like him, have transformed their soils and agricultural productivity. Elias's own cereal yields are up fourfold, and this agricultural success has boosted the local economy, with families moving back from the capital, Tegucigalpa. The demand for labour has put wages at close to double those in nearby villages. All children now finish their primary schooling, and seven from Pacayas have gone on to secondary school. Elias's own daughter is now a teacher at the local school. A neighbour of Elias says: *'Now, no one ever talks of leaving.'* People are more content with their own place, and they can choose from a range of futures.

Further west of here lies another transformed farm in the village of Guacamayas, which belongs to Irma de Guittierez Mendez. It, too, is in the hills – in fact, 85 per cent of Honduras is located on slopes that are steeper than 15 per cent. Irma farms on the edge of La Tigra National Park, the *watershed* for the capital city's drinking water. Her farm is another model for farmers everywhere – she, too, works with nature rather than battles against it. The farm is covered with *terracita*, small terraces to conserve soil and water. She grows maize; cassava; and four beans; seven vegetables; banana; guava and avocado; and coffee under apple trees at the top of the slope. These crops are rotated in order to control diseases, and Irma brings wasp nests from the forest to hang on the farm trees, which control pests. She makes her own compost and buys in chicken manures.

Importantly, Irma is also a teacher, both of fellow farmers and of professional agronomists who come to the valley to see this revolution for their own eyes. She says: *'One of the things we were taught was the responsibility of anyone who knows something to teach it to others in the community. As a result, we can think more about what we are doing now. Community spirit has improved.'* Perhaps some may find this a curious attitude in a world where modern and competitive methods of agriculture dominate. But Irma is modest: *'Our purpose is not to make a lot of money, but to help the community as a whole.'* There is also a bigger picture to these improvements. As farmers find ways to improve the quality and health of their own soils, so the likelihood of them illegally moving onto the national park is drastically reduced. This makes the park authority happy because they can spend less on guards and weapons. Like all these cases of agricultural transformation, not everything is perfect. Farmers struggle to find markets, infrastructure is poor, and research and extension agencies are often unaware of the progress that is being made.

Nonetheless, Irma and Elias are not alone in forging a new way for agriculture for those who have been excluded. One of the most extraordinary changes to have occurred in the past decade and a half is the quiet emergence of a revolution in developing country agriculture amongst small and large farmers. It has been driven by farmers who are increasingly rich in knowledge about nature, and how to use it to increase food production, and who have the willingness and capacity to collaborate in order to solve common problems. Many poorer farmers have made the transition from largely pre-modern agricultural systems directly to sustainable and highly productive systems. In every case, there are important lessons for us all.

## Critical Choices for Agricultural Development

The gloomy global predictions about increasing numbers of people, growing demand for cereals and meat, and stubbornly persistent hunger and poverty raise an important question: whom should we target? Many now agree that if women have access to food, and the means to produce it, then this food is also more likely to get into the mouths of children. Low birth weight is now known to be a vital factor in both child malnutrition and premature death. But this, in turn, is caused by a mother's poor nutrition before conception and during pregnancy. Foetal under-nutrition also contributes to increasing incidences of chronic disease in later life. Each year, 30 million infants are born in developing countries with impaired growth, comprising 6 per cent of children in South-East Asia and Latin America, and close to 15 per cent in sub-Saharan Africa. In the year 2000, a quarter of pre-school children in developing countries had stunted growth, where height is less than two standard deviations for the age, with proportions rising to 50 per cent in East and South-Central Asia.[1]

But this stunting carries forward to school-age children, too. In four out of ten countries recently surveyed, more than one third of children were stunted. This is due to insufficient and poor-quality food, with deficiencies of key vitamins and minerals most common. Worldwide, 2 billion people suffer from iron-deficiency anaemia, including three-quarters of pregnant women in South-East Asia, half in Africa, one third in the Americas, and one quarter in Europe. Anaemia causes 65,000 maternal deaths per year in Asia, and severe vitamin A deficiency affects 100 million to 250 million children worldwide.[2]

What can be done? Lisa Smith and Lawrence Haddad's review of the past 25 years of child malnutrition suggests that improved food availability is only one of four factors that are important in overcoming child health problems. The other three are improved female education, access to family health services, and status improvements for women relative to men. Women are disadvantaged in agricultural systems. They produce up to 80 per cent of food, but have access to less than 10 per cent of credit and extension advice, and also own very little land. The United Nations' fourth report on the world's nutrition says:

> *Investing in maternal and childhood nutrition will have both short- and long-term benefits of huge economic and social significance, including reduced health care costs throughout the life cycle, increased educability and intellectual capacity, and increased adult productivity. No economic analysis can fully capture the benefits of such sustained mental, physical and social development.*[3]

It is clear that an adequate and appropriate food supply is a necessary condition for eliminating hunger and food poverty. But increased food supply does not automatically mean increased *food security* for all. What is important is who produces the food, who has access to the technology and knowledge to produce it, and who has the purchasing power to acquire it. The conventional wisdom is that, in order to double food supply, we need to redouble efforts to modernize agriculture. After all, it has been successful in the past. But there are major doubts about the capacity of such systems to produce the food where the poor and hungry people live. These people need low-cost and readily available technologies to increase food production. A further challenge is that this must happen without further damage to the environment.

All of this leaves us with three choices for agricultural development if we are to increase food supply. We could expand the area of agriculture by converting new lands to crop and animal production, but with the result that important services from forests, grasslands and other areas of important biodiversity are lost. We could increase per *hectare* production in agricultural exporting countries, mostly industrialized, so that food can be transferred or sold to those who need it, but with the result that the poor will continue to be excluded. Or we could seek to increase total farm productivity in developing countries that most need the food.

We know the first two choices work in terms of increased food production. The third has also worked for farmers with access to sufficient inputs of pesticides, fertilizers and other modern technologies. But most farmers in developing countries are not in such a position. If they are poor, or the country is poor, then there is no option of purchasing inputs in order to increase productivity. This is before considering whether such approaches might or might not cause harm to the environment or to human health. The success of modern agriculture in recent decades has often masked damaging *externalities* or side effects, and it is only recently that large-scale environmental and health costs have come to be appreciated.[4] Thus, the specific question we need to ask is: to what extent can farmers improve food production with cheap, low-cost, locally available technologies and inputs that do not cause harm to the environment or to human health when used?

What, then, do we understand by sustainable agriculture? In the first instance, a more sustainable farming seeks to make the best use of nature's goods and services while not damaging the environment.[5] It does this by integrating natural processes such as nutrient cycling, nitrogen fixation, soil regeneration and natural enemies of pests into food production processes. It also minimizes the use of non-renewable inputs that damage the environment or harm the health of farmers and consumers. It makes use of the knowledge and skills of farmers, thereby improving their self-reliance, and it seeks to make productive use of people's collective capacities to work together in order to solve common management problems, such as pest, watershed, irrigation, forest and credit management.

Sustainable agriculture is also *multifunctional* within landscapes and economies. It jointly produces food and other goods for farm families and markets; but it also contributes to a range of *public goods*, such as clean water, biodiversity, *carbon sequestration* in soils, groundwater recharge, flood protection, and landscape amenity value. As sustainable agriculture also seeks to make the best use of nature, so technologies and practices must be locally adapted. They are most likely to emerge from new configurations of social relations (comprising relations of trust embodied in new social organizations) and new horizontal and vertical partnerships between institutions, and from human capacity (comprising leadership, ingenuity, management skills and the capacity to innovate). Thus, agricultural systems with high levels of social and human assets are more able to innovate in the face of uncertainty.

## Does Sustainable Agriculture Work?

These are all fine ideas, but can they work in practice? At the University of Essex, we recently completed the largest survey of sustainable agriculture improvements in developing countries. The aim was to audit progress towards agricultural sustainability, and assess the extent to which such initiatives, if spread on a much larger scale, could feed a growing world population that is already substantially food insecure.[6] We looked at more than 200 projects in 52 countries, including 45 in Latin America, 63 in Asia and 100 in Africa.[7] We calculate that almost 9 million farmers were using sustainable agriculture practices on about 29 million hectares, more than 98 per cent of which emerged in the past decade.[8] These methods are working particularly well for small farmers; about half of those surveyed are in projects with a mean area per farmer of less than 1 hectare, and 90 per cent are in areas with less than 2 hectares each.[9]

We found that improvements in food production are occurring through one or more of four different mechanisms. The first involves the intensification of a single component of farm system, with little change to the rest of the farm, such as home garden intensification with vegetables and/or tree crops, vegetables on rice bunds, and the introduction of fish ponds or a dairy cow. The second involves the addition of a new productive element to a farm system, such as fish or shrimps in paddy rice, or agroforestry, which provides a boost to total farm food production and/or income, but which does not necessarily affect cereal productivity. The third involves better use of nature to increase total farm production, especially water (by water harvesting and irrigation scheduling) and land (by reclamation of degraded land). This leads to additional new dryland crops and/or increased supply of additional water for irrigated crops, and thus increases cropping intensity. The fourth involves improvements in per hectare yields of staples through the introduction of new regenerative elements into farm systems, such as legumes and *integrated pest management*, and new and locally appropriate crop varieties and animal breeds.

As a result, a successful sustainable agriculture project may substantially improve domestic food consumption or may increase local food barters or sales through home gardens or fish in rice fields, or better water management, without necessarily affecting the per hectare yields of cereals. Home garden intensification occurred in one fifth of projects; but given its small scale, it accounted for less than 1 per cent of the area. Better use of land and water, giving rise to increased cropping intensity, occurred in one seventh of projects, with one third of farmers and one twelfth of the area. The incorporation of new productive elements within farm systems, mainly fish and shrimps in paddy rice, occurred in 4 per cent of projects, and accounted for the smallest proportion of farmers and area. The most common mechanisms comprised yield improvements with regenerative technologies or new seeds/breeds, which occurred in 60 per cent of the projects, with an uptake of more than half of the farmers and about 90 per cent of the area.

What is happening to food production? We found that sustainable agriculture has led to an average 93 per cent increase in per hectare food production. The relative yield increases are greater at lower yields, indicating greater benefits for poor farmers and for those who have not benefited from the recent decades of modern agricultural development.[10] The increases are quite remarkable, as most agriculturalists would be satisfied with any technology that can increase annual productivity by even 1 or 2 per cent. It is worth restating: these projects are seeing close to a doubling of per hectare productivity over several years, and this still underestimates the additional benefits of intensive food production in small patches of home gardens or fish ponds.

We also calculated the increase in food production for those projects with reliable data on yields, area and numbers of farmers. In the 80 projects with less than 5-hectare farms where cereals were the main staples, 4.5 million farmers on 3.5 million hectares increased household food production by more than 1.5 tonnes per year, an increase of 73 per cent. In the 14 projects with potato, sweet potato and cassava as the main staples, the 146,000 farmers increased household food production by 17 tonnes per year, an increase of 150 per cent. In the projects in southern Latin America with a larger farm size (an average of 90 hectares per farm), farm production increased by 150 tonnes per year, an increase of 46 per cent.

These aggregate figures understate the benefits of increased diversity in the diet, as well as increased quantity. Most of these agricultural sustainability initiatives have seen increases in farm diversity. In many cases, this translates into increased diversity of food consumed by the household, including fish protein from rice fields or fish ponds, milk and animal products from dairy cows, poultry and pigs kept in the home garden, and vegetables and fruit from home gardens and farm micro-environments. Although these initiatives are reporting significant increases in food production, some as yield improvements, and some as increases in cropping intensity or diversity of produce, few are reporting surpluses of food being sold to local markets. This is because of a significant elasticity of consumption amongst rural households experiencing any degree of food insecurity. As production increases, so also does domestic consumption, with direct benefit, in particular, for women's and children's health. In short, rural people are eating more food and a greater diversity of food, and this does not show up in the international statistics.

I acknowledge that all of this may sound too good to be true for those who would disbelieve these advances. Many still believe that food production and nature must be separated, that *agroecological* approaches offer only marginal opportunities to increase food production, and that industrialized approaches represent the best, and perhaps only, way forward. However, prevailing views have changed substantially in just the last decade, and many sceptics are beginning to recognize the value and innovative capacity emerging from poorer communities in developing countries.

There are four types of technical improvements that play substantial roles in these agroecological food-production increases: soil health improvements; more efficient water use in both dryland and irrigated farming; pest and weed control with minimum or zero-pesticide use; and whole-system redesigns. In each, there are many stories of new thinking and innovative practices. Once again, I cannot do these examples justice by telling the whole story. Nor is there the space to dwell on specific difficulties and limitations. This agricultural sustainability revolution is not one thing – it is comprised of many elements that are adapted to localities and are, inevitably, different from place to place. By telling these stories and cases, I therefore do not want to imply that the same approaches and technologies will work everywhere. What is important, though, are the principles of collective action, locally adapted science and innovation, and making the best of what nature can offer through agroecological approaches to food production.

## Soil Health Improvements

Soil health is fundamental for agricultural sustainability. It is the most important part of any agricultural system – the fundamental asset. When soils are in poor health, they cannot maintain productive agriculture. Yet, today, many agricultural systems are under threat because soils have been damaged, eroded or simply ignored during the process of agricultural intensification.[11] It is estimated that nearly 2 billion hectares of land worldwide are degraded. They suffer from a mix of physical degradation by water and wind erosion, crusting, sealing and waterlogging; chemical degradation by acidification, nutrient depletion, pollution from industrial wastes and excessive use of pesticides and fertilizers; and biological degradation by organic matter depletion, and loss of soil flora and fauna.[12] Three-quarters of degraded land is in Africa (490 million hectares), Asia (750 million hectares) and Latin America (240 million hectares), with Europe, North America and Australia each having 100 million to 200 million hectares degraded. In Africa, farmland is annually losing nitrogen, phosphate and potassium nutrients at a rate of at least 30 kilogrammes per hectare, with land in 23 countries losing more than 60 kilogrammes per hectare.[13]

Sustainable agriculture starts with the soil by seeking to reduce soil erosion and to make improvements to soil physical structure, organic matter content, water-holding capacity and nutrient balances. Soil health is improved through the use of legumes, green manures and cover crops; the incorporation of plants with the capacity to release phosphate from the soil into rotations; the use of composts and animal manures; the adoption of *zero-tillage*; and the use of inorganic fertilizers where needed.[14] Some of these are age-old practices adapted for today's conditions. Some, though, seem to break one of the fundamental rules of agriculture. Ever since the birth of farming some 12,000 years ago, farmers have been ploughing, or tilling, the soil. Yet, in the past decade, Latin American farmers have found that eliminating tillage can be highly beneficial, and many in Africa have adopted no-till or only shallow cultivations for rice production. At first sight, it seems a strange idea. After harvest, the crop residues are left on the surface to protect against erosion. At planting, seed is slotted into a groove that is cut into the soil. Weeds are controlled with herbicides or cover crops. This means that the soil surface is always covered, and the soil itself no longer inverted.

The fastest uptake of these minimum till systems has been in Brazil, where there are some 15 million hectares under *plantio direto* (also called zero-tillage even though there is some disturbance of the soil), mostly in three southern states of Santa Caterina, Rio Grande do Sul and Paraná, and in the central Cerrado. In neighbouring Argentina, there are more than 11 million hectares under zero-tillage, up from less than 100,000 hectares in 1990, and in Paraguay there are another 1 million hectares of zero-tillage.[15] There are several million hectares of conservation or no-till farming in the US, Canada and Australia; but here it mostly tends to be simplified modern agriculture systems, which save on soil erosion but do not necessarily make the best use of agroecological principles for nutrient, weed and pest management.

In Argentina, Roberto Peiretti is responsible for technical and management decisions on about 10,000 hectares of farmland in Córdoba Province. He is chairman of the Argentinian no-till farmers' organization and is an enthusiast. He says:

> We were faced with serious soil deterioration, and knew we needed to find a different way to produce. . . applying no till as an entirely holistic approach enabled us to discover an entirely new scenery, a system based on understanding and emulating nature as much as possible.

Their approach has been to establish no-till research and extension groups, and to link these to regional and national levels. These coalitions have been vital to the rapid spread of no-till farming. The system clearly works. In Argentina, average cereal productivity was 2 tonnes per hectare in 1990; since then, it has increased by about 10 per cent on conventional farms, a rate far surpassed by those farms with zero-tillage, where yields have doubled. On Roberto's farm, the 2001 harvest has been the best to date. From his field, he says: '*I am busy but happy because we are again able to have higher yields for our soybean, corn and sorghum. The oldest no-till paddocks are peaking at nearly five tonnes per hectare for soybean, ten tonnes for corn, and eight tonnes for sorghum.*' I asked him what he was most proud of. He says: '*The land has become fertile. We clearly see the wildlife has increased in our farms, there is water in the soil, and farmers are better off. I feel that there is strong correlation between feeling well, and being conscious of living within a framework of environmentally friendly attitudes.*'

To the north in Brazil, the transformations in the landscape and in farmers' attitudes are equally impressive. John Landers runs a network of Clubes Amigos da Terra, friends of the land clubs, in the Cerrado, the vast area of formerly unproductive lands colonized for farming during the past two decades. These lands needed lime and phosphorus before they could become productive. He believes that zero-tillage represents '*a total change in the values of how to plant crops and manage soils. On adopting zero-tillage, farmers adopt a higher level of management and become environmentally responsible.*' There are many fundamental changes, including '*the adoption of biological controls, awareness that the new technology is eliminating erosion and building the soil so they have something to leave for their children, and a willingness to participate in joint actions.*'

Zero-tillage has had an effect on social systems, as well as on soils. In the early days, there was a widespread belief that zero-tillage was only for large farmers. This has now changed, and small farmers are benefiting from technology breakthroughs developed for mechanical farming. A core element of zero-tillage adoption in South America has been adaptive research – working with farmers at microcatchment level to ensure technologies are fitted well to local circumstances. According to Landers: '*Zero-tillage has been a major factor in changing the top-down nature of agricultural services to farmers towards a participatory, on-farm approach.*' There are many types of farmers' groups: from local (farmer microcatchment and credit groups), to municipal (soil commissions, Friends of Land clubs, commercial farmers' and farm workers' unions), to multimunicipal (farmer foundations and cooperatives), to river basin (basin committees for all water users), and to state and national level (state zero-tillage associations and the national zero-tillage federation).

Farmers are now adapting technologies – organic matter levels have improved so much that fertilizer use has been reduced and rainfall infiltration improved. Farmers are now getting rid of contour terraces at many locations, insisting that there are no erosion problems. As biological controls are enhanced with surface mulch and crop rotations, it has also become possible to reduce the amount of pesticides used, with some success in herbicide-free management. Other benefits of zero-tillage include reduced siltation of reservoirs, less flooding, higher aquifer recharge, lowered costs of water treatment, cleaner rivers, and more winter feed for wild biodiversity.[16] A large public good is also being created when soil health is improved with increased organic matter. Organic matter contains carbon, and it is now recognized that soils can act as sites for carbon sequestration. Not only are these sustainable agriculture farmers creating a soil of good health, they are providing a benefit to us all by sequestering large amounts of carbon from the atmosphere, in this way mitigating the effects of climate change. However, there is still controversy over zero-tillage. Some feel that the use of herbicides to control weeds, or the use of genetically modified crops, means that we cannot call these systems sustainable. However, the environmental benefits are substantial, and new research is already showing that farmers have effective agro-ecological alternatives, particularly if they use cover crops for green manures in order to raise organic matter levels.[17]

In the Sahelian countries of Africa, the major constraints to food production are also related to soils, most of which are sandy and low in organic matter. In Senegal, soil erosion and degradation threaten large areas of agricultural land; and since the late 1980s, the Rodale Institute Regenerative Agriculture Resource Centre has worked closely with farmers' associations and government researchers to improve the quality of soils. The primary cropping system of the region is a millet-groundnut rotation. Fields are cleared by burning, and then cultivated with shallow tillage using animals. But fallow periods have decreased dramatically, and inorganic fertilizers do not return high yields unless there are concurrent improvements in organic matter. Soils low in organic matter also do not retain moisture well.

The Rodale Centre now works with about 2000 farmers who are organized into 59 groups on improving soil quality by integrating stall-fed livestock into crop systems, by adding legumes and green manures, by increasing the use of manures, composts and rock phosphate, and by developing water-harvesting systems. The result has been a 75 to 190 per cent improvement in millet and groundnut yields – from about 300 to about 600–900 kilogrammes per hectare. Yields are also less variable year on year, with consequent improvements in household food security. Amadou Diop summarizes an important lesson for us all: *'Crop yields are ultimately uncoupled from annual rainfall amounts. Droughts, while having a negative effect on yields, now do not result in total crop failure.'*[18]

This is the critical message – improve the soil, and the whole agricultural system's health improves, too. Even if this is done on a very small scale, people can benefit substantially. In Kenya, the Association for Better Land Husbandry found that farmers who constructed double-dug beds in their gardens could produce enough vegetables to see them through the hungry dry season. These raised beds are improved with composts, and green and animal manures. A considerable investment in labour is required; but the better water holding capacity and higher organic matter means that these beds are both more productive and better able to sustain vegetable growth through the dry season. Once this investment is made, little more has to be done for the next two to three years. Women, in particular, are cultivating many vegetable and fruit crops, including kale, onion, tomato, cabbage, passion fruit, pigeon pea, spinach, pepper, green bean and soya. According to one review of 26 communities, three-quarters of participating households are now free from hunger during the year, and the proportion having to buy vegetables has fallen from 85 to 11 per cent.[19]

For too long, agriculturalists have been sceptical about these organic and conservation methods. They say they need too much labour, are too traditional, and have no impact on the rest of the farm. Yet, you only have to speak to the women involved to find out what a difference they can make. In Kakamega, Joyce Odari has 12 raised beds on her farm. They are so productive that she now employs four young men from the village. She says: *'If you could do your whole farm with organic approaches, then I'd be a millionaire. The money now comes looking for me.'* She is also aware of the wider benefits: *'My aim is to conserve the forest, because the forest gives us rain. When we work our farms, we don't need to go to the forest. This farming will protect me and my community, as people now know they can feed themselves.'* Once again, the spin-off benefits are substantial – giving women the means to improve their food production means that food gets into the mouths of children. They suffer fewer months of hunger, and so are less likely to miss school.[20]

## Improved Water Efficiencies

The proper management of water is also essential for agriculture. Too much or too little, and crops and animals die. Carefully managed, though, and landscapes become productive. About one fifth of the world's cropland is irrigated, allowing food to be produced in dry seasons when rainfall is in short supply but sunlight is abundant. In some parts of the tropics, farmers produce three crops each year, and altogether irrigated lands produce two-fifths of the world's food. Most farmers, though, are entirely dependent on rainfall, an input that is becoming increasingly erratic and uncertain in the face of climate change.

Despite a long history of agricultural systems and cultures built upon complex water management, ranging from the irrigated rice cultures of Asia to Roman cultivation of Libyan North Africa, from irrigated Mesopotamia and Egypt to the flood-water farming cultures of the Papago, Hopi and Navajo peoples of the American south-west, water as a common resource is still under-valued and under-managed today. There is great scope for improvement; once again, farmers in many developing countries are leading the way. Through better social organization, they are finding that shared management and cooperation can lead to greater returns for whole systems.

Water harvesting has a wide application in the drylands. In northern India, in the uplands of Gujarat, Rajasthan and Madhya Pradesh, land degradation is severe, soils are poor, and agricultural production is so low that most families need someone working in the city in order to survive. Again, with the right approach and best sustainable practices, much is being done. The Indo-British Rainfed Farming project, for example, works with 230 local groups in 70 villages on water harvesting, tree planting, and land grazing improvements. Basic grain yields of rice, wheat, pigeon peas and sorghum have increased from 400, to 800, to 1000 kilogrammes per hectare, and the increased fodder grass production from the terrace bunds is highly valued for livestock. The improved water retention has resulted in water tables rising by 1 metre over three to four years, meaning that an extra crop is now possible for many farmers, thus turning an unproductive season into a productive one.

Women are again the major beneficiaries. P S Sodhi of the local group Gram Vikas Trust in Udaipur puts it this way:

*In these regions, women never had seen themselves at the front edge of doing things, taking decisions, and dealing with financial transactions. The learning by doing approach of the project has given them much needed confidence, skills, importance and awareness.*

The wider benefits of a transformed agriculture are also evident:

*The project has indirectly affected migration as people are gaining more income locally through the various enterprises carried out in the project. People are now thinking that they must diversify more into new strategies. There has also been a decline in drawing on resources from the forests.*

But perhaps more importantly:

*People have also started to question the nature of democratic participation. They have also started to challenge the political systems — those who are in power or control power have little incentive to allow participatory institutions to develop. Yet in our villages, people are voicing their concerns, have overruled elites, and have even elected women as Sarpanchs, local leaders.*

In sub-Saharan Africa, water harvesting is also turning barren lands green. Again, the technologies are not complex and costly, and can be used in even the poorest of communities. In central Burkina Faso, 100,000 hectares of abandoned and degraded lands have been restored with the adoption of *tassas* and *zaï*. These are 20- to 30-centimetre holes dug in soils that have been sealed by a surface layer hardened by wind and water erosion. The holes are filled with manure to promote termite activity and to enhance infiltration. When it rains, water is channelled by simple stone bunds to the holes, which fill with water, and into which are planted millet or sorghum seeds. Normally, cereal yields in these regions are precariously low, rarely exceeding 300 kilogrammes per hectare. Yet, these lands now produce 700 to 1000 kilogrammes per hectare. Chris Reij of the Free University in Amsterdam found that the average family in Burkina Faso who used these technologies had shifted from being in annual cereal deficit amounting nearly to 650 kilogrammes, equivalent to six and a half months of food shortage, to producing a surplus of 150 kilogrammes a year. *Tassas* are best suited to landholdings where family labour is available, or where farm labour can be hired. The soil and water conservation methods have led to a market for young day labourers who, rather than migrate, now earn money by building these structures.[21]

Good organization helps to improve irrigated agriculture, too. Despite great investment, many irrigation systems have become inefficient and subject to persistent conflict. Irrigation engineers assume that they know best how to distribute water, yet can never know enough about the specific conditions and needs of large numbers of farmers. Recent years, though, have seen the spread of a very simple idea — help organize farmers into water users' groups, and let them manage the water distribution for themselves. One of the best examples comes from the Gal Oya region in Sri Lanka. Before this approach, Gal Oya was the largest and most run-down scheme in the country. Now, farmers' groups manage water for 26,000 hectares of rice fields, and produce more rice crops per year and per unit of water. Moreover, when farmers took control, the number of complaints received by the irrigation department about water distribution fell to nearly zero.

The benefits were dramatically shown during the 1998 drought. According to the government, there was only enough water for the irrigation of 18 per cent of the rice area. But farmers persuaded the irrigation department to let this water through on the grounds that they would carefully irrigate the whole area. Through cooperation and careful management, they achieved a better than average harvest, earning the country US$20 million in foreign exchange.[22] Throughout Sri Lanka, 33,000 water users' associations have now been formed – a dramatic increase in local social organization that has improved farmers' own capacities for problem-solving and cooperation, and for using nature more efficiently and effectively in order to produce more food.

## Zero-Pesticide Farming

Modern farmers have come to depend upon a great variety of insecticides, herbicides and fungicides to control the pests, weeds and diseases that threaten crop and animal productivity. These pesticides are now big business, with global sales exceeding US$31 billion in 1998. Each year, farmers apply 5 billion kilogrammes of pesticides' active ingredients to their farms. Nine-tenths of this market is now controlled by just eight companies. Yet, it is only in the past century, less than 1 per cent of agriculture's history, that such dependence has emerged.[23] Today, however, many farmers in this agricultural sustainability revolution are finding alternative methods for pest, disease and weed control. In some crops, it may mean the end of pesticides altogether, as cheaper and more environmentally benign practices are found to be perfectly effective.

Though integrated pest management dates back to the 1950s, a significant paradigm-shifting moment occurred in the early 1980s when Peter Kenmore and his colleagues in South-East Asia counter-intuitively found that pest attack on rice was directly proportional to the amount of pesticides used. In other words, more pesticides meant more pests. The reason was simple – pesticides were killing the natural enemies of insect pests, such as spiders and beetles. When these invertebrates are eliminated from agroecosystems, then pests are able to expand rapidly in numbers. This led, in 1986, to the banning by the Indonesian government of 57 types of pesticides for use on rice, combined with the launching of a national system of farmer field schools to teach farmers the benefits of biodiversity in fields. One million farmers have now attended about 50,000 field schools, the largest number in any Asian country. The outcomes in terms of human and social development have been remarkable, and farmer field schools are now being deployed in many parts of the world. Agriculturalists now believe that irrigated rice can, for most of the time, be grown without pesticides, provided the biodiversity is present.

Many countries are now reporting large reductions in pesticide use. In Vietnam, 2 million farmers have cut pesticide use from more than three sprays to one per season; in Sri Lanka, 55,000 farmers have reduced use from three to one half sprays per season; and in Indonesia, 1 million farmers have cut use from three sprays to one per season. In no case has reduced pesticide use led to lower rice yields.[24] Amongst these are reports that many farmers are now able to grow rice entirely without pesticides: one quarter of field-school trained farmers in Indonesia, one fifth to one third in the Mekong Delta of Vietnam, and three-quarters in parts of the Philippines.

The key to success is biological diversity on farms. Pests and diseases like monocultures and *monoscapes* because there is an abundance of food and no natural enemies to check their growth. In the end, they have no fear of pesticides, as resistance inevitably develops within populations and spreads rapidly unless farmers are able to keep using new products. Moreover, when a harmful element is removed from an agricultural system, and biodiversity is managed to provide free pest-management services, then further options for redesign are possible. Traditionally, rice paddies were important sources of fish protein, and fish living in fields helped in nutrient cycling and pest control. But pesticides are toxic to fish, and their increased use since the 1960s entirely eliminated beneficial fish from paddies. Take the pesticides away, though, and the fish can be reintroduced.

In Bangladesh, a combined aquaculture and integrated pest management programme is being implemented by CARE with the support of the UK government and the European Union.[25] Six thousand farmer field schools have been completed, with 150,000 farmers adopting more sustainable rice production on about 50,000 hectares. The programmes also emphasize fish cultivation in paddy fields and vegetable cultivation on rice field dykes. Rice yields have improved by about 5–7 per cent, and costs of production have fallen owing to reduced pesticide use. Each hectare of paddy, though, yields up to 750 kilogrammes of fish, an extraordinary increase in total system productivity for poor farmers with very few resources. Farmers themselves recognize the changes in farm biodiversity. One said to Tim Robertson, former leader of the programme: *'Our fields are singing again, after 30 years of silence.'* It is the frogs singing in diverse and healthy fields that are full of fish and rice. Arif Rashid of CARE estimates that 85,000 farmers have stopped using insecticides; but as he says: *'We do not know how much this has spread to other farmers.'* I asked him what he thought was the most significant element of success, and it is clear again that farmers have crossed a frontier:

> *CARE was able to change the behaviour of participating farmers with regard to irrational use of fertilizers and unwise use of insecticides, and they now have an improved understanding of ecology. They now take decisions based on careful study of their farms.*

Once we start with the idea that diverse systems can provide enough food, particularly for farmers with few resources, then whole new fields of scientific endeavour can emerge. One of these is the science of *semio-chemicals*, aromatic compounds given off by plants. In East Africa, Hans Herren, winner of the World Food Prize for work on a parasite to control the cassava mealybug, is director of the International Centre for Insect Physiology and Ecology. He believes that minimum- to zero-pesticide farming systems are possible throughout the tropics, and his centre is researching sustainable pest management through biological control, using one organism to control another, botanical agents for natural pest-control compounds derived from plants, habitat management, and pest-tolerant varieties of food crops.

In Kenya, researchers from the International Centre for Insect Physiology and Ecology (ICIPE) and Rothamsted in the UK found that maize produces semiochemicals when fed upon by the stem borer (*Chilo* spp.). They also found that these same chemicals increase foraging and attack by parasitic wasps, and are fortuitously also released by a variety of local grasses used for livestock fodder and soil erosion control.[26] The interactions are complex. Napier and sudan grass attract stem borers to lay their eggs on the grass instead of the maize. Another grass, molasses grass, and a legume, *Desmodium*, repel stem borers. Both napier and molasses grass emit another chemical that summons the borers' natural enemies, so that pest meets predator. There is yet more, as *Desmodium* not only fixes nitrogen but is allelopathic (toxic) to the parasitic witchweed *Striga hermonthica*.

Researchers call their redesigned and diverse maize fields *vutu sukumu*: push–pull in Swahili. They clearly work, as more than 2000 farmers in western Kenya have adopted maize, grass-strip and legume-intercropping systems, and have at the same time increased maize yields by 60 to 70 cent. The sad truth is that for 30 years, the official advice to maize growers in the tropics has been to create monocultures for modern varieties of maize, and then apply pesticide and fertilizers to make them productive. Yet, this agricultural simplification eliminated vital and free pest management services produced by the grasses and legumes. *Vutu sukumu* systems are complex and diverse, and they are cheap – they do not rely on costly purchased inputs.[27]

## Whole System Synergies

What we do not yet know is whether a transition to sustainable agriculture (delivering greater benefits at the scale occurring in these projects) will result in enough food to meet the current food needs in developing countries, let alone the future needs after continued population growth and the adoption of more urban and meat-rich diets. But what we are seeing is highly promising. There is also scope for additional confidence, as evidence indicates that productivity can grow over time if natural, social and human assets are accumulated. These findings are similar to those of Jeff McNeely and Sara Scherr, whose recent review of ecoagriculture in both developing and industrialized countries has also indicated that there are novel ways in which to feed the world and to save biodiversity.[28]

The issue of asset accumulation over time is important. If agricultural systems are low in natural, social and human assets (either intrinsically low or damaged by degradation), then a sudden switch to 'more sustainable' practices that rely on these very assets will not be immediately successful – or, at least, not as successful as it might be. In Cuba, for example, urban organic gardens produced 4000 tonnes of food in 1994. Over just five years, production grew to more than 700,000 tonnes, partly because of an increase in the number of gardens, but also because the per area productivity had steadily risen over time.[29]

Increased productivity over time has been found in fish ponds in Malawi. These are tiny, typically 200 to 500 square metres in size, and are integrated within a farm so that they recycle wastes from other agricultural and household enterprises. In 1990, yields were 800 kilogrammes per hectare, but rose steadily to nearly 1500 kilogrammes per hectare over six years. Randy Brummet of the International Centre for Aquatic Resource Management indicates why: *As farmers gain a greater understanding of how this new system functions, and an appreciation of its potential, they become increasingly able to guide further evolution towards increasing productivity and profitability.*[30] Revealingly, when non-participatory approaches were used to work with farmers, and systems were imposed in a completed format on farmers, then yields fell.

Each type of sustainable agriculture improvement, by itself, can make a positive contribution to raising production. But another dividend comes with combinations. Synergistic effects tend not to be captured or appreciated by reductionist methods of analysis that measure the effects of one variable at a time, while holding all the others unchanged – the so-called *ceteris paribus* approach. But this ignores synergism – where the whole is greater than the sum of the parts. Thus, soil and water conservation that emphasizes terracing and other physical measures to prevent soil loss is much less effective than combinations of biological methods that attempt to increase the productivity of the system combined with finance for credit groups that reduces the indebtedness of households.

Sustainable agriculture systems become more productive when human capacity increases, particularly farmers' capacity to innovate and adapt their farm systems for sustainable outcomes. Sustainable agriculture is not a concretely defined set of technologies, nor is it a simple model or package to be widely applied or fixed with time. It needs to be conceived of as a process for social learning. Lack of information on *agroecology* and the necessary skills to manage complex farms is a major barrier to adopting sustainable agriculture. We know much less about these resource-conserving technologies than we do about the use of external inputs in modernized systems. So, it is clear that the process through which farmers learn about technology alternatives is crucial. If they are enforced or coerced, then they may only adopt for a limited period. But if the process is participatory and enhances farmers' ecological literacy of their farms and resources, then the foundation for redesign and continuous innovation is laid. As Roland Bunch and Gabino Lòpez have put it about Central American agriculture: *'What needs to be made sustainable is the social process of innovation itself.'*

## Madagascar's System of Rice Intensification

I have already talked of the low-pesticide and high social-connectivity revolutions in rice management. Another revolution may be about to emerge from remote and impoverished Madagascar. It is called the System of Rice Intensification (SRI), and it breaks many of the rules of rice cultivation developed over thousands of years. It was first developed by Father Henri de Laudanié during the 1980s, and has been tested and promoted by the local Association Tefy Saina, with the help of Norman Uphoff and colleagues at Cornell University. The system has improved rice yields from about 2 tonnes per hectare to 5, 10 and even 15 tonnes per hectare on farmers' fields. This has been achieved without using purchased inputs of pesticides or fertilizers. The improvements have been so extraordinary that, until lately, they have been disbelieved and ignored by most scientists. SRI challenges so many of the basic principles of irrigated rice cultivation, and produces such extraordinary productivity increases, that most professionals have elected to be sceptical.

The system is centred on making better use of the existing genetic potential of rice. Firstly, rice seedlings are transplanted after 8 to 12 days, instead of the normal 30 to 50 days. Early transplanting increases tillering, and SRI plants typically have 50 to 80 tillers, compared with 5 to 20 for conventional ones. Each tiller bears a head of grain. Secondly, rice seedlings are usually planted close together in order to minimize weed infestation. But in the SRI, they are planted at least 25 centimetres apart in a grid pattern rather than in rows. This facilitates mechanical weeding, as well as saving on costly seed – the system uses about 7 rather than 100 kilogrammes of seed per hectare. More widely spaced plants develop a different architecture, with more room for roots and tillers; and better root systems mean reduced lodging (ie, the likelihood of stem weakness and collapse).

Most scientists and farmers believe that rice, as an aquatic plant, grows best in standing water. In the SRI, however, paddies are kept unflooded during the period of vegetative growth. Water is only applied to keep the soil moist, which is allowed to dry out for periods of three to six days. Only after flowering are paddies flooded. They are then drained 25 days before harvest, as with conventional rice. Such management encourages more root growth. Since flooding is the conventional approach to weed control, SRI farmers must weed up to four times – mechanically or by hand. Farmers who do not weed still get respectable yield increases of twofold to threefold. But those who weed get increases of fourfold to sixfold. SRI farmers also use compost rather than inorganic fertilizers.

The proof that SRI works comes from the number of farmers using it – an estimated 20,000 farmers have now adopted the full SRI, and Sebastien Rafaralahy of Tefy Saina estimates that another 50,000 to 100,000 farmers are now experimenting with elements of the system. Cornell scientists led by Norman Uphoff have helped research institutions in China, Indonesia, the Philippines, Cambodia, Nepal, the Côte d'Ivoire, Sri Lanka, Cuba, Sierra Leone and Bangladesh to test SRI. In all cases, severalfold jumps in rice yields were achieved. In China, for example, yields of 9 to 10 tonnes per hectare were achieved in the first year, compared with a national average of 6 tonnes.[31]

## Salinity Farming in Vietnam

It takes a sharp eye and an open mind to see the possibilities in complex systems. Vo-Tong Xuan of Angiang University has both of these, and he saw something special 20 years ago on the Ca Mau Peninsula of southern Vietnam. Here, saline water and problem soils have been turned to farmers' advantage by adopting rice-shrimp systems combined with novel methods of water and soil management. The peninsula has no sources of freshwater except rain and very deep wells, but it has mangroves, fields and an abundant marine environment. On a field trip with a group of students, Xuan stopped at Long Dien Dong village to show a soil profile. Pointing to the bluish-grey soil horizon below the brown topsoil, he told farmer Ba Sen: *'If you let this soil dry, it will become acidic, and nothing can be grown on it. But when the field is permanently wet, even with saline water on the surface, your soil fertility will be maintained forever.'* But to Ba Sen, this was not news at all. He had accidentally discovered a sustainable practice of managing the potentially acid sulphate soil four years before, and had even written it in a will to his eldest son.

The rice field is prepared during the start of the rainy season. Saline water is let out of the field and the soil is flushed by the rainwater. Initially, soil salinity might be high, but it falls after a few rains. Seedlings are prepared in nurseries in the early rainy season, and fields are cleaned of weeds and algae, without tillage. Seedlings are transplanted at the age of 30 to 40 days at the end of July, and Ba Sen gets a yield of about 4 tonnes per hectare. After harvest, and while the soil is still wet and river water not yet saline, farmers allow river water to enter the field to raise shrimps.[32] The first shrimp harvest is after the end of the rainy season, after which the field is prepared for another crop of shrimp. This time, saline water is taken into the field at high tide. Water management is crucial, and water is exchanged once or twice a week in order to create a continuous flow in the field. Stocking of shrimps is carried out between January and March, and these are fed with cassava, coconut meal, milled rice and fishmeal, yielding (during April to June) nearly 200 kilogrammes per hectare.

What is clever about this system is that Ba Sen is getting much more from rather unpromising resources than anyone could have expected. He has spoken at local meetings and international conferences, and the practices have spread throughout Ca Mau. The result is that mangroves are being sustained, marine resources valued, and agriculture's productivity increased – all because an integrated and balanced system of management has been developed in which the total is better than the sum of its parts. As Vo-Tong Xuan puts it:

> *The creative and intelligent people of Ca Mau now have a rich experience in exploiting their saline water environment. They do not see it as a constraint to their development, but on the contrary they take advantage of it, a valuable advantage which will lead Ca Mau to prosperity.*

# Ecological Reconstruction in China

Bei Guan village lies in the rolling hills and plains of Yanqing County, under the shadow of the Great Wall of China. It is the site of a remarkable experiment in integrating sustainable agriculture with renewable energy production. Bei Guan was selected by the ministry of agriculture as an ecological demonstration village for implementing integrated farming systems in one of 150 counties across the country.[33] It has made the transition from monocultural maize cultivation to diverse vegetable, pig and poultry production. Each of the 350 households has a tiny plot of land, about 2 *mu* (one seventh of a hectare), a pen for the livestock, and a biogas digester. Ten types of vegetable are grown and sold directly into Beijing markets. The green wastes are fed to the animals, and their wastes are channelled into the digester. This produces methane gas for cooking, lighting and heating, and the solids from the digester are used to fertilize the soil. Each farmer also uses plastic sheeting to create greenhouses from the end of August to May, thus extending production through the biting winter when temperatures regularly fall to minus 30 degrees Centigrade.

The benefits for local people and the environment are substantial – more income from the vegetables, better and more diverse food, reduced costs for fertilizers, reduced workload for women, and better living conditions in the house and kitchen. In Bei Guan, there is also a straw gasification plant that uses only maize husks to produce gas in order to supplement household production. Instead of burning husks in inefficient stoves, requiring 500 baskets per day for the whole village, just 20 are burned per day in the plant. The village head, Lei Zheng Kuan, says: *'These have saved us a lot of time. Before, women had to rush back from the fields to collect wood or husks, and if it had been raining, the whole house would be full of smoke. Now it is so clean and easy.'*

The benefits of these systems are far reaching. The ministry of agriculture promotes a variety of integrated models across the country, involving mixtures of biogas digesters, fruit and vegetable gardens, underground water tanks, solar greenhouses, solar stoves and heaters, and pigs and poultry. These are fitted to local conditions. As Wang Jiuchen, director of the ministry's division of renewable energy, says: *'If farmers do not participate in this ecological reconstruction, it will not work.'* Whole integrated systems are now being demonstrated across many regions of China, and altogether 8.5 million households have digesters. The target for the coming decade is the construction of another 1 million digesters per year. Because the systems of waste digestion and energy production substitute for fuel wood, coal or inefficient crop-residue burning, the benefits for the natural environment are substantial – each digester saves the equivalent of 1.5

tonnes of wood per year, or 3 to 5 *mu* of forest. Each year, these biogas digesters are effectively preventing 6–7 million tonnes of carbon from being emitted to the atmosphere, a substantial benefit to us all.

## Confounding Factors and Trade-Offs

This agricultural sustainability revolution is clearly benefiting poor people and environments in developing countries. People have more food, are better organized, are able to access external services and power structures, and have more choices in their lives. But like all major changes, this revolution may also provoke secondary problems. For example, building a road near a forest can help farmers to reach food markets, but may also aid illegal timber extraction. This is not to say that depletion of natural assets is always undesirable. It may be in the national and local interests to convert part of a forest into finance, if that money is to be used for investment in hospitals and schools, effectively producing a transfer from natural to social and human capital. Equally, short-term social conflict may be necessary for overcoming inequitable land ownership in order to produce better welfare outcomes for the majority. Projects may make considerable progress in reducing soil erosion and in increasing water conservation through the adoption of zero-tillage, but may still continue to rely on applications of herbicides. In other cases, improved organic matter levels in soils may lead to increased leaching of nitrate to groundwater. If land has to be closed off to grazing for rehabilitation, then people with no other source of feed may have to sell their livestock; and if cropping intensity increases or new lands are taken into cultivation, then the burden of increased workloads may fall on women, in particular. Additional incomes arising from sales of produce may also go directly to men in households, who are less likely than women to invest in children and the household as a whole.

There are also a variety of emergent factors that could slow the spread of sustainable agriculture. Firstly, sustainable agriculture that increases the asset base may simply increase the incentives for more powerful interests to take over, such as landlords taking back formerly degraded land from tenants who had adopted sustainable agriculture. In these contexts, it is rational for farmers to farm badly – at least they get to keep the land. The idea of sustainable agriculture may also appear to be keeping people in rural areas away from centres of power and from 'modern' urban society; yet, some rural people's aspirations may precisely be to gain sufficient resources to leave rural areas. Sustainable agriculture also implies a limited role for agrochemical companies, as currently configured – and these companies will not accept such market losses lightly. Sustainable agriculture, furthermore, suggests greater decentralization of power to local communities and groups, combined with more local decision-making, both of which might be opposed by those who benefit from corruption and non-transparency in private and public organizations. Research and extension agencies will have to change, too – adopting more participatory approaches in order to work closely with farmers. These agencies must adopt different measures to evaluate job success and reasons for promotion. Finally, social connectivity, relations of trust, and the emergence of significant movements may represent a threat to existing power bases, who, in turn, may seek to undermine such locally based institutions.

There will be many who actively dispute the evidence of promising success, believing that the poor and marginalized cannot possibly make these kinds of improvements. But I believe that there is great hope and leadership in these stories of progress towards agricultural sustainability. What is quite clear is that they offer real opportunities for people to improve their food production while protecting and improving nature. Sustainability will be difficult to achieve on a wide scale because many individuals will oppose these ideas, dismiss the innovators, or resist policy reforms. Yet, here lie some pointers to salvation, if we all could but listen and learn.

## Concluding Comments

Food poverty remains a daily challenge for more than 800 million people, despite great progress with industrialized agriculture. Hunger accompanies increased food productivity. We know how to increase food production with modern methods and fossil-fuel derived inputs; but anything that costs money inevitably puts it out of reach of the poorest households and countries. Sustainable agriculture, in seeking to make the best use of nature's goods and services, combined with people's own capacities for collective action, offers many new opportunities. There has already been great progress, though sceptics remain unconvinced that the poor can be innovative. Sustainable agricultural systems improve soil health, increase water efficiency and make the best use of biodiversity for the control of pests and diseases. When put together, there are important synergistic interactions that improve the system's performance as a whole. Sadly, there remain many confounding factors that will make wider adoption of, and transition to, sustainable agriculture difficult without substantial policy reform.

## Chapter 5

# *Only Reconnect*

## Such Great Success. . .

Let me take you for a summer's walk through one of today's temperate wheat fields. As we tread carefully, pushing through the stiff, golden stalks, nothing seems to stir. Between each plant the soil is bare and hard. On the horizon, a smudge of green suggests a field boundary, perhaps a distant hedgerow or lonely tree. We are now standing on a factory floor, at the centre of an efficient machine that does little else but produce food. Like any factory, it does a job well, but it has no place for nature. The same goes for most animal raising. In North America, cattle once spent their whole lives roaming the prairies. But today, they are packed into feedlots, as many as 100,000 animals at a time. They are efficiently fed, put on 1 kilogramme or more of meat each day, and together produce wastes equivalent to those from a sizeable city. These, too, are food factories, and many say they represent great progress in efficient food production. But is this true?

One of the rarest animals in the world persists in patches of these industrialized landscapes. It is the Suffolk Punch, the giant horse first bred in the 16th century to work the heavy Suffolk clays of eastern England. Suffolks are tall, often with a white star or blaze on the face, and have long been admired for their calm temperament and ease of care.[1] But in the modern era, such shire horses could not compete with machinery, and since the 1950s they were rapidly replaced with tractors and mechanized combines. Farms, of course, became more efficient. More land was cultivated in less time with less labour. But when these horses and their horsemen disappeared from farms, something else was lost, too. The horsemen had an intimate relationship not just with their horses, but with the whole farm landscape. They were expert botanists, using up to 40 species of wild plants for horse care. Today, having forgotten this knowledge, we call these once useful plants weeds, and the Suffolk only survives through the efforts of dedicated societies and individuals, one or two of whom still farm with shire horses.

From generation to generation, horsemen passed on knowledge about the value of certain plants for treating illness and disease, shining the coat or improving appetite. George Ewart Evans, eloquent observer of English agricultural change, wrote in *The Horse and the Furrow* of fevers treated with agrimony or with apples sliced and stored until infested with antibiotic-carrying fungi; and of colds and coughs cured with fever few, belladonna, meadow-rue and horehound. For de-worming, the horsemen used celandine, yellow-flowered indicator of spring, and to encourage appetite, put gentian, elecampine, horehound and felwort into food. They used box to keep down sweat, and burdock, saffron, rosemary, fennel, juniper, tansy and mandrake for coat conditioning. Hazel, holly and willow were fashioned into withies and traces for harnesses. This example shows that there is a simple principle for our modern era of agricultural progress. As food efficiency increases, so landscape diversity is lost, and so, too, goes an intimate knowledge of nature and a duty of care.[2]

Far from the clays of Suffolk, Kevin Niemeyer stands in the shade of his veranda, looking out on one of the most fertile landscapes of Queensland. This is the Lockyer Valley, sub-tropical vegetable garden of eastern Australia, and home to another modern revolution in nature-friendly farming. In a land farmed for only a few generations, crisis point was reached during the late 1980s and 1990s. Every two or three days, Kevin had to spray his brassicas with pesticides. But this pattern of use carried an ecological hazard, as pests quickly developed resistance. When Kevin bought his farm in the 1970s, he did not need to spray much for the first few years. Later, he found he had to spray more often as the beneficial insects disappeared. Thus, the seeds of failure are contained within a modern agriculture that has to kill nature in order to survive. Over the years, the situation worsened. On the verge of quitting farming because of pesticide resistance (nothing seemed to work any more), Kevin was asked by Sue Heisswolf of the local research station to try something different. The aim was to develop a system dependent upon natural pest control methods.[3] The psychological barriers to overcome were massive. Kevin says: *'I was called a nut but I had a go. I put all my crops on the line, and eventually the people who called me a nut came back and asked me how I did it.'*

Sue and Kevin later helped to form the Brassica Improvement Group to bring together 30 or so local farmers to experiment with new farm methods and to share their results. They began regular scouting for pests, cut conventional pesticide use and adhered to a summer production break. They introduced predators, pheromone strips, and natural products, such as *Bacillus thuringiensis* (*B.t.*) sprays, and manipulated the farm habitat by adding trees to encourage birds and planting allysum in cabbage rows to provide food for beneficial insects. The impact has been startling. Says Kevin: *'Crops which would have been sprayed 36 times in three months are now only sprayed once or twice with a natural pesticide.'* The fields are now full of green frogs, wasps, spiders and birds, all providing a free service in the form of pest control.

Many others in the valley have got the message, too, and aggregate pesticide use has fallen dramatically. But not all farmers have changed. When I asked the group what was their biggest worry, they said *'our fathers'*, who kept on asking *'when will you go back to farming properly, rather than messing around with these strange methods'*. Just 500 metres from Kevin's farm, a neighbour continues to spray every two days, even though Kevin's farmland biodiversity has done the job perfectly for the past ten weeks with no need for any intervention. His broccoli performs best, as he has not sprayed that for three years. Kevin reflects on this fundamental challenge for redesigning ecological and social landscape: *'Conventional farming has played havoc with our farms, but farmers still have difficulty changing.'* And change they all must, for the forces of ecological and economic change are stacking up.

## Commodities or Culture?

Modern agricultural methods provoked a 50-year revolution in farming in industrialized countries. They brought spectacular increases in productivity – more cereals and animals per *hectare*, more meat and milk per animal, more food output per person employed. The fear of widespread hunger has largely been banished, as productivity has grown in almost every sector. In the UK, wheat yields remained largely unchanged from the 1880s to the 1940s, around 2–2.5 tonnes per hectare. Since then, there has been a rapid increase to reach an average of 8 tonnes per hectare today.[4] In the US, each dairy cow produces nearly 8000 kilogrammes of milk each year, more than triple that of a cow 50 years ago. Over the same period, beef cattle have increased in size by 22 per cent, pigs by 90 per cent, and broiler chickens by 52 per cent.[5]

At the same time, the scale of production has grown. Small farmers have been swallowed up, and large operators have thrived and expanded even further. The industry has become bigger and better at producing food as a commodity, most of which is now grown or reared in massive monocultures. Whereas once farming was based on mixed enterprises, with livestock wastes returned to the land, and cereal and vegetable by-products fed to animals, now enterprises are increasingly specialized and geographically separated. Should we be concerned about these losses of cultural diversity? Or should we resist any attempt to see farming as anything other than an efficient producer of the commodities that we all need on a daily basis?

One of the most striking changes has been the growth in scale of livestock farming, and the shift towards confined systems that rely entirely on imported feedstuffs. The trend has been the same in every industrialized country – but the effect has been the greatest in the US. Huge livestock operations have emerged in the pig, dairy, broiler and layer chicken, and beef sectors. For many of these enterprises, it is no longer correct to use the term 'farm'.[6] In Colorado and Texas, five companies own 27 feedlots on which 1.5 million cattle are penned, an average of 60,000 animals per feedlot. A single feedlot of 240 hectares in California contains 100,000 animals, finishing more than 200,000 each year. Four hundred animals are squeezed into each hectare, and each animal puts on about 1.5 kilogrammes daily, staying in the feedlot for four to five months. As they consume about 10 kilogrammes of feed each day, there is a great deal of waste. A feedlot this size produces 100,000 tonnes of waste per year, and uses 4 million litres of water a day in the summer. Just to top it all, the beef is sold under the company's own brand name as 'ranch beef', evoking days of open prairies and traditional cowboy culture.[7]

This growth and skewing of the size of farm operations is mirrored by growing concentration, at every stage, in the food chain. There are fewer input suppliers, fewer farms, fewer millers, slaughterers and packing businesses, and fewer processors. Increasingly, one business owns a whole piece of the food chain, producing the feed, raising the livestock, slaughtering and packing them, and then selling the products to consumers in their own shops. Bill Heffernan and colleagues at the University of Missouri have been tracking the concentration ratio of the top four firms in various food sectors for many years. Today in the US, the largest four firms control 79 per cent of beef packing and 57 per cent of pork packing. The concentration ratio for broiler and turkey producers is between 40–50 per cent. For flour milling, dry and wet corn milling, and soybean crushing, the ratio varies between 57–80 per cent.[8]

Big scale also brings simplification and loss of biological diversity. Worldwide, there are 6500 breeds of domesticated animals and birds, including cattle, goats, sheep, buffalo, yaks, pigs, horses, chicken, turkeys, ducks, geese, pigeons and ostriches. One third of these are under immediate threat of extinction owing to their very small population size. Over the past century, it is believed that 5000 animal and bird breeds have already been lost. The situation is most serious in the already industrialized farming systems, with half of breeds at risk in Europe and one third at risk in North America. We must now worry that those countries currently with fewer breeds at risk, 10–20 per cent in Asia, Africa and Latin America, will follow the same route as the industrialized countries.[9]

For some, such large-scale operations and loss of diversity are a measure of success. Food commodity prices have been falling steadily over the past 20 years, and most industrialized countries have moved well away from the threat of food shortages. It was only in 1954, after all, that the UK ended food rationing. However, in this success lies the seed of destruction. Large-scale industrialized farming looks good precisely because it measures its own success narrowly and ignores the costly side effects.[10] There are many signs that our highly productive and modernized systems are now in crisis. Farmers have been dispossessed, food and environmental safety compromised, and food insecurity allowed to persist. Consumers are increasingly disconnected from the process of food production, and disenchantment grows. Aldo Leopold, perceptive observer of our relations of nature, saw the changes coming more than half a century ago when he said: *'If the individual has a warm personal understanding of the land, he will perceive of his own accord that it is something more than a breadbasket.'*[11]

## The End of the Family Farm Culture?

The realities of industrialized farming contrast painfully with the pastoral notions of an agricultural system that many consumers still hold dear. Rural communities are dying all over the industrialized world; but the food system appears to go from strength to strength. In an old landscape only recently converted to farming, yet another farm sale takes place in the morning mist. In the Mid-West grain bowl of North America, and home to generations of family farmers, the gavel smacks down on piles of rusting machinery, mournful animals, and acres of desolate farmland. The life and history of another farm family is dispersed to the four winds. The farm is swallowed up, so that another farm can compete better, until that, too, needs to get bigger again. During the past 50 years, 4 million farms have disappeared in the US. This is equivalent to 219 for every single one of those 18,000 days.[12]

In France, 9 million farms in 1880 became just 1.5 million by the 1990s. In Japan, 6 million farmers in 1950 became 4 million by 2000. Many advocates of economic progress and efficiency say that these are predictable and perhaps sad losses, but inevitable if we are to have progress. Farmers increase their productivity, the inefficient are weeded out, and the remaining farms are better able to compete on world markets.

But each of these lost farm families used to have a close connection with the land, and to other farms in their communities. When they are disconnected, the memories are lost forever. Strangely, again, we call this progress. John Steinbeck saw this coming more than 60 years ago in *The Grapes of Wrath*, when he lamented:

*And when a horse stops work and goes into the barn there is life and a vitality left, there is breathing and a warmth, and the feet shift on the straw, and the jaws clamp on the hay, and the ears and the eyes are alive. There is a warmth of life in the barn, and the heat and smell of life. But when the motor of a tractor stops, it is as dead as the ore it came from. The heat goes out of it like the living heat that leaves a corpse. Then the corrugated iron doors are closed and the tractor man drives home to town, perhaps twenty miles away, and he need not come back for weeks or months, for the tractor is dead. And this is easy and efficient. So easy that the wonder goes out of the work, so efficient that the wonder goes out of the land and the working of it, and with the wonder the deep understanding and the relation.[13]*

In the US, the changing numbers of farmers and average farm size show an interesting pattern. Farm numbers increased steadily from 1.5 million to more than 6 million from 1860 to the 1920s, as the frontier was pushed back, stabilized for 30 years, but then fell rapidly since the 1950s to today's 2 million. Over the same period, average farm size remained remarkably stable for 100 years, around 60–80 hectares; but it climbed from the 1950s to today's average of 187 hectares.[14]

However, hidden in these averages are deeply worrying trends. Only 4 per cent of all US farms are over 800 hectares in size, and 47 per cent are smaller than 40 hectares. Technically, 94 per cent of US farms are defined as small farms – but they receive only 41 per cent of all farm receipts. Thus, 120,000 farms out of the total of 2 million receive 60 per cent of all income. The recent National Commission on Small Farms noted: 'The pace of industrialization of agriculture has quickened. The dominant trend is a few large vertically integrated farms controlling the majority of food and fibre products in an increasingly global processing and distribution system.'[15]

Tom Dobbs of South Dakota State University, in his evidence to the National Commission, describes what happened in eastern South Dakota, where his great grandfather first set up a farm in the 1870s.[16] Three generations were raised on the farm, and it finally passed out of family ownership in 1997. In Moody County, the location of the farm, farm numbers halved from 1300 in 1949 to 640 during the 1990s, with size doubling to 180 hectares. But it is in the standardization of the landscape where change has been most dramatic. Soybean acreage rose sharply, and areas under oats, flax, hay and barley fell, accompanied by large falls in numbers of sheep and small declines in cattle and pigs. Mixed systems were replaced by simplified systems of maize and soybean. As Tom Dobbs says, these changes mirror those across the Corn Belt and Great Plains, with small farms replaced by large farms, and mixed farms by simple ones.

It is only narrow economics that allows us to believe that these large operations are actually more efficient. We simply do not use the proper accounting measures. The National Commission also indicated that: 'Normal measures of efficiency do not reflect the social and environmental goods produced by a large number of small farms.' Willis Peterson of the University of Minnesota echoes this sentiment by asserting: 'The small family and part-time farms are at least as efficient as larger commercial enterprises. In fact, there is evidence of diseconomies of scale as farm size increases.'[17]

In two previous books, *Regenerating Agriculture* and *The Living Land*, I have reflected on the historic analysis by Walter Goldschmidt of California in the mid 20th century. It bears restating, in brief. Goldschmidt studied the two communities of Arvin and Dinuba in the San Joaquin Valley, similar in all respects except for farm size. Dinuba was characterized by small family farms, and Arvin by large corporate enterprises. The impact of these structures of farming was remarkable. In Dinuba, he found a better quality of life, superior public services and facilities, more parks, shops and retail trade, twice the number of organizations for civic and social improvement, and better participation by the public. The small farm community was seen as a better place to live because, as Michael Perelman later put it: 'The small farm offered the opportunity for "attachment" to local culture and care for the surrounding land.' A study 30 years later confirmed these findings – social connectedness, trust and participation in community life was greater where farm scale was smaller.[18]

Yet, small farmers are still widely viewed as economically inefficient. Their disappearance has, in truth, been a severe loss to rural culture. Linda Lobao's study of rural inequality shows the importance of locality that

Goldschmidt's research illustrated. The decline of family farming does not just harm farmers. It hurts quality of life in the whole of society. Corporate farms are good for productivity, but not much else. They bring a decline in rural population, increased poverty and income inequality, lowered numbers of community services, diminished democratic participation, decreased retail trade, and increased environmental pollution. Says Lobao: *This type of farming is very limited in what it can do for a community. . . we need farms that will be viable in the future, correspond to local needs and remain wedded to the community.*[19]

Wendell Berry, poet and farmer, has long drawn attention to what happens during modernization. An agricultural crisis, he says, is a crisis of culture:

> *A healthy farm culture can be based only upon familiarity and can grow only among people soundly established on the land; it nourishes and safeguards a human intelligence of the earth that no amount of technology can satisfactorily replace. The growth of such a culture was once a strong possibility in the farm communities of this country. We now have only the sad remnants of those communities. If we allow another generation to pass without doing what is necessary to enhance and embolden the possibility now perishing with them, we will lose it altogether.*[20]

Nevertheless, since Berry wrote this in the 1970s, another generation has passed.

The central question is really this: are we content with agricultural systems becoming larger in scale and producing anonymous commodities, or do we expect something more from them?[21] Family farms do more than just produce food. They help to build a tangible culture of connections to the land. Lorraine Garkovich and her colleagues' study of farm families in Kentucky shows how important are the accumulated connections between family farms and the land: *The family farm is more than just soil and livestock. It is also traditional strategies for how to farm, care for, and use the land and traditional meanings and values attached to the land.*[22] On these farms, time passes slowly and experience accumulates into individual and collective memories. These farmers are good at story-telling, and these stories bind communities, giving meaning and direction to lives. But when the shared understanding breaks down, then dissatisfaction and, eventually, conflict can emerge. Today, family farmers mourn the decline of rural communities; no one has time to talk anymore, and many people in rural areas no longer know anything about farming. Canadian author Sharon Butala says:

> *The most potent reason of all to save small-scale family farms [is] because those who farmed in this way had the time to ponder and enjoy and be instructed and*

*inspired by nature. When there's no one left out there except people whose days in the land are spent twenty feet off the ground in air-conditioned cabs of tractors. . . who will remember how to be on the land? Who will remember how to listen to the land?*[23]

With the increasing scale and centralization of modern agriculture have come the standardization and simplification of whole landscapes. Landscapes with diversity have many functions and niches, but *monoscapes* are poor performers.[24] They have lost vital ecological functions, and therefore are less resilient. Put simply, monoscapes do one thing well, and that is produce abundant food. But they are fundamentally unhealthy and disconnected systems of food production. They arise because certain individuals are able to claim the common benefits of landscape for themselves, with few checks and balances. *Diverscapes* do so much more – they jointly produce food, support people's livelihoods, and preserve nature as a result of economic activity, not as a sideline. Diverscapes are *multifunctional* and polycultural, full of uncertainty, mysteries and difference. As farmer and writer David Kline has put it: *'I believe we need some unconcreted mysteries. We need the delight of the unknown and the unexplainable in nature.'*[25] The only mystery in modern farming is that we have failed to understand the associated environmental and health costs.

One sad result of the *modernist* project is increasing place neutrality. This is a beguiling vision, as it appears to offer independence, the ability to come and go as you please, without reference to the cloying and parochial ties of places and localities. At the same time, though, such place neutrality implies all the facilities but none of the heart, none of the natural connectedness between people, and between people and land.[26] In this modern world, should we bother about those who say they have important connections to a place? Why not just let market forces dictate, and gather up our belongings and move whenever necessary? Today, it takes the same amount of time to fly half way around the world as it did a century ago to travel 50 kilometres. So why not pull up these roots and make the best of opportunity?

There is one simple reason, and that is because of our desire to have a place we can call home. Home, as Deborah Tall puts it, *'is where we know – and are known – through accumulated experience'.*[27] It is not something that happens quickly. It gives us stability and meaning. It is where we have best established connections with people and with place. It is where we return after long journeys. It is where the food on our plate has some local identity. For all these reasons, it is not a commodity to be traded easily. People relocated from condemned slums often suffer badly because they feel that they have lost their real home. Each year, one fifth of all

Americans move house, meaning that an average American moves 14 times in a lifetime. Homes have become commodities, and you trade up when you can at the cost of losing some sense of permanence. Even worse, this promotes a distrust of those who do remain in one place for a lifetime, who are, as Tall puts it: *'often interpreted as being unambitious, unadventurous – a negation of American values'*. More worryingly, the meaning of place is changing. It is increasingly something that is centrally designed and created, rather than accumulated over time. Inevitably, this means increasing disconnection from local distinctiveness and nature, resulting in places coming *'to mean proximity to highways, shopping and year-round recreation, rather than natural situation or indigenous character'*.[28]

## Countering the Shrinking Food Pound

At the turn of the 21st century, farming cultures are now in crisis all over the industrialized world. How can this be? How can an industry showing extraordinary growth in productivity, and sustained over decades, have lost public confidence owing to persistent environmental damage and growing food safety concerns? The food that is supposed to sustain us is now a source of ill health for many, and the systems that produce that food damage the environment. This can no longer be right.

Once again, the devil is in the detail. One of the reasons why many farmers struggle is that the proportion of the food pound or dollar that is returned to farmers has shrunk. Fifty years ago, farmers in Europe and North America received as income between 45–60 per cent of the money consumers spent on food. Today, that proportion has dropped dramatically to just 7 per cent in the UK and 3–4 per cent in the US, though it remains at 18 per cent in France.[29] So, even though the global food sector continues to expand, now standing at US$1.5 trillion a year, farmers are getting a relatively smaller share. Over time, the value of food has been increasingly captured by manufacturers, processors and retailers. Farmers sell the basic commodity, and others add the value. As a result, less money gets back to rural communities and cultures; and they, in turn, suffer economic decline. A typical wheat farmer, for example, receives 6 cents of each dollar spent on bread, about the same as for the wrapping. But if farmers are receiving such a small proportion of the food pound and dollar, what happens when they sell directly to consumers? Do their farms and landscapes change for the better?

Jan and Tim Deane were the first farmers in the UK to sell vegetables directly to local consumers through a formalized box scheme. Their 12-hectare smallholding in Devon would barely register as a field on a conventional large farm. Yet, they grow 60 types of vegetables, and supply them fresh to 200 customers each week in a boxed selection. It all started in 1984, when they bought Northwood Farm on land not well suited to market gardening. Says Jan:

> *We had the usual disasters in those first years – pest problems, weed problems and, especially, 15 years ago, difficulties in finding suitable markets. Together with several other growers from Cornwall to Hampshire, we were founding members of an organic marketing coop that sold to retail shops, the wholesale market and the supermarkets. But by the end of 1990, it was obvious to us that we were never going to survive financially growing 10 or 12 acres of organic vegetables for the pre-pack and wholesale markets. We were too small, the land too indifferent, and as producers in the south-west, we were too far away from the major markets so that transport availability and cost was an ongoing headache and financial drain.*

Their cooperative became a casualty, and they began packing a small range of vegetables to order in a desperate bid for extra income. They contacted existing customers and other friends and neighbours whom they felt might be interested and offered to pack and deliver a box of mixed vegetables each week for a fixed price. The response was encouraging and they began to pack 20 boxes a week. These were so popular that within two years they had dispensed with the wholesale market altogether and made the box system the sole means of distribution and income generation.

Through occasional questionnaires and casual conversation as they delivered the boxes, they gradually built up a detailed picture of what their customers wanted in the boxes. They produced a newsletter once a month so that they could inform their customers about what was happening on the farm and at what stages the crops were. This helped customers to increase their awareness of what is actually involved in growing vegetables and helped them to feel part of something special. On the annual Northwood Farm walk, customers had the opportunity to spend an afternoon with Tim and Jan, who took them on conducted tours of what they were going to eat in the coming months. Jan says:

> *It was a revelation to us to realize how divorced so many people are from the realities of food production. Some of our customers were shocked to realize that we could actually grow vegetables in fields – in their mind's eye, Northwood was just a rather larger than normal garden – and the reality took some adjustment. However, when it came to rabbits we never really managed to convince them that the cute little bunnies that they saw hopping away in the distance were to us both a nuisance and a potential threat to our survival.*

The vegetable season lasts 32 weeks from June to February or March. Over the years, Tim and Jan have increased the number of species grown from 20 to 60, and found that the farm seemed to benefit from this. Among other things there seemed to be better stocks of predators to control pests. Moreover, all customers live within 6 kilometres of the farm. Most importantly for the Deanes, they receive all of the food pound, and they make more money this way. Jan again says: *'We more than doubled the profitability of the farm and for the first time found ourselves on a secure financial footing.'* Northwood Farm has gone on to become the prototype of the box and community-supported agriculture schemes that have developed so successfully in the UK during the last decade.

## Systems of Sustainable Agriculture

One reason for this sharply falling share of the food pound is modern farming's increased dependence on purchased inputs and technologies. In the latter part of the 20th century, external inputs of pesticides, inorganic fertilizer, animal feedstuffs, energy, tractors and other machinery have become the main means to increase food production. These external inputs, though, have substituted for free natural-control processes and resources, rendering them more vulnerable. Pesticides have replaced biological, cultural and mechanical methods of controlling pests, weeds and diseases. Inorganic fertilizers have substituted for livestock manures, composts, nitrogen-fixing crops and fertile soils. What were once valued local resources have all too often become waste products. These changes would represent a major problem if alternatives did not exist. Now they do. Sustainable agriculture technologies do two important things. They conserve existing on-farm resources, such as nutrients, predators, water or soil; and they introduce new elements into the farming system that add to the stocks of these resources, such as nitrogen-fixing crops, water-harvesting structures or new predators. These then substitute for some or all of the external inputs.

Many of the individual technologies are multifunctional, and their adoption results, simultaneously, in favourable changes in several aspects of farm systems. For example, hedgerows encourage wildlife and predators and act as windbreaks, thereby reducing soil erosion. Legumes in rotations fix nitrogen, and also act as a break crop to prevent carry-over of pests and diseases. Clovers in pastures reduce fertilizer bills and lift sward digestibility for cattle. Grass contour strips slow surface run-off of water, encourage percolation to groundwater, and are a source of fodder for livestock. Catch crops prevent soil erosion and leaching during critical periods, and can also be ploughed in as a green manure. Green manures not only provide a readily available source of nutrients for the growing crop, but also increase soil organic matter and hence water retentive capacity, further reducing susceptibility to erosion. Low-lying grasslands that are managed as water meadows, and that provide habitats for wildlife, also provide an early-season yield of grass for lambs.

In Europe, about one third of the farmed landscape, some 56 million hectares, is still under relatively unintensive agricultural systems.[30] These traditional systems are typically highly diverse, and are closely linked to particular ways of life for rural communities. They include the high mountain pastures of southern and central Europe, characterized by transhumance and summer migrations of livestock; the valley farms of the Carpathians, almost entirely under traditional management, with hay meadows rich in flowering plants; the diverse wood-pasture systems of Portugal and Spain (the *montados* and *dehesas*), characterized by species-rich grasslands, mixtures of cork and holm oaks, sheep, pigs and cattle, and a remarkable abundance of rare birds; the 3–4 million hectares of traditionally managed fruit trees, olives and vines in Greece, Italy, Spain and Portugal that are also good for wildlife; the mixed farms of Hungary, Poland and Ireland; and the wet and dry grasslands of France and Italy. Sadly, almost all of these systems are under severe threat from both modern farming methods and rural abandonment.

In contrast to these traditional systems, organic farming represents a deliberate attempt to make the best use of local natural resources. The aim of organic farming, also known as ecological or biological agriculture, is to create integrated, humane, and environmentally and economically viable agriculture systems in which maximum reliance is put on locally or farm-derived renewable resources, and the management of ecological and biological processes. The use of external inputs, whether inorganic or organic, is reduced as far as possible. Recent years have seen a dramatic increase in the adoption of organic farming. In Europe, the extent has increased from just 100,000 hectares in 1985 to more than 3 million hectares, managed by 120,000 farmers in 2000. In the US, 550,000 hectares of land managed by 5000 growers were certified under organic production in 1997. The important thing for most organic farmers is that it represents a system of agriculture rather than simply a set of technologies.[31] The primary aim is to find ways in which to grow food in harmony with nature. The term organic, as Nic Lampkin of the Welsh Institute for Rural Studies has put it, is *'best thought of as referring not to the type of inputs used, but to the concept of the farm as an organism, in which the component parts — the soil minerals, organic matter, micro-organisms, insects, plants, animals and humans — interact to create a coherent and stable whole'.*[32]

These interconnections are important. Lady Eve Balfour, founder of the Soil Association, author of the 1940s book *The Living Soil*, and owner of an experimental farm at Haughley in Suffolk, saw agriculture as a vital service for the nation: *'If the nation's health depends on the way its food is grown, then agriculture must be looked upon as one of the health services, in fact the primary health service.'* She, like other founders of the organic movement, Albert Howard and Friend Sykes, saw agriculture as intimately connected with human and environmental health. This should change the way we think about food production: *'Once agriculture comes to be regarded as a health service, the only consideration in any matter concerning the production of food would be is it necessary for the health of the people? That of ordinary economics would take a quite secondary place.'*[83]

Another type of agricultural sustainability in industrialized landscapes is represented by what has been called integrated farming. This is another environmentally friendly approach to farming. Once again, the emphasis is upon integrating technologies to produce site-specific management systems for whole farms, incorporating a higher input of management and information for planning, setting targets and monitoring progress.

There are important historical, financial and policy reasons why still relatively few farmers have taken the leap from modernist farming to organic agriculture. But it is possible for anyone to take a small step that can, in theory, be followed by another step. Integrated farming in its various guises represents a step or several steps towards sustainability. What has become increasingly clear is that modern farming is wasteful, as integrated farmers have found that they can cut purchased inputs without losing out on profitability. Some of these cuts in use are substantial; others are relatively small. By adopting better targeting and precision methods, there is less wastage and therefore more benefit to the environment. Farmers can then make greater cuts in input use once they substitute some regenerative technologies for external inputs, such as legumes for inorganic fertilizers or predators for pesticides. Finally, they can replace some or all of the external inputs entirely over time once they have fully adopted, and learned about, a new type of farming characterized by new goals and technologies.[34]

## Bioregional Connections to Sustainable Foodsheds

The basic challenge for a more sustainable agriculture is to make best use of available natural and social resources. Farming does not have to produce its food by damaging or destroying the environment. Farms can be productive and farmers earn a decent living while protecting the landscape and its natural resources for future generations. Farming does not have to be dislocated from local rural cultures. Sustainable agriculture, with its need for increased knowledge, management skills and labour, offers new upstream and downstream job opportunities for businesses and people in rural areas. This suggests a logical need to emphasize agriculture's connections to local ecologies and communities.

In this modern world, those of us who are not farmers express our connections with nature in combinations of three ways – by visiting it, by joining organizations, and by eating the food. Firstly, we visit and observe it, walk in it, bathe in it, occasionally at weekends or on annual holidays, sometimes daily while walking the dog.[35] Each year in the UK, we make more than 550 million day and overnight visits to the countryside and seaside, spending a total of UK£14 billion in local economies. This is more than four times as great an input to rural areas as subsidies to farmers from government. The choices that we make on these visits thus make a big difference to the supply of goods and services, whether directly in the form of food, or indirectly in the form of landscapes.[36]

We also join organizations which we feel are engaged in activities to protect, conserve or regenerate those aspects of nature or countryside that we value. Environmental, heritage and countryside organizations have now become some of the largest membership organizations in industrialized countries. In aggregate, they have overtaken political party membership, and are second only to trades unions. These represent a wide range of different voices, pulling in many different directions.[37] Many began as protest movements, and then later evolved to take on a more positive 'solutions-oriented' agenda. The economic and political powers of these organizations come from the membership base. In the UK, both the Royal Society for the Protection of Birds (with more than 1 million members) and the Wildlife Trusts (more than 300,000 members) now own large amounts of land, both reserves and farms, and are demonstrating that positive management can make a difference. The largest landowner in the UK after the crown is the National Trust, which owns 275,000 hectares and has 2.5 million members. In the US, the Sierra Club has 600,000 members, the National Audubon Society 550,000 members, and the Wilderness Society 200,000 members.

Perhaps most importantly, because it is a daily activity, we eat the food produced from the farms that shape nature on a daily basis. We vote once every two, three or four years, yet we shop every week, or even every day. We must have food, and in having it we also encourage the system of production that brought it from land to larder. This means that the food system as a whole deserves to be described as another commons. It is something that belongs to us all. Yet, in an unrestrained or unregulated context, the tragedy is that we over-consume and under-invest in this commons. Worse, we appear not to appreciate the consequences.

When food is a commodity, there is little to stop over-consumption. There are no checks and balances to have us worry about the hidden costs of certain types of food production. Our current food system, despite considerable performance improvements in recent decades (it is faster, fitter and more streamlined), is still flawed. It simply is not working to the advantage of its 6 billion commoners. There is hunger at the tables of 800 million people. At the same time, there is widespread obesity. This cannot be right; nevertheless, by our action, it already is accepted. However, collective action by producers of food, by consumers, and by novel mixtures of both groups can make a difference. It is possible to create new forms of relationship, trust and understanding, leading to new cognitive constructions of food and its cultures of production.

Two concepts are useful in this rethinking – the ideas of bioregions and *foodsheds*. *Bioregionalism* implies the integration of human activities within ecological limits, and bioregions are seen as diverse areas with many ecological functions. Bioregionalism can thus be seen as a self-organizing or *autopoiëtic* concept that connects social and natural systems at a place people can call home. Bioregions are real places where people want to live. They take years to build, emerging from the interactions of people who are not indifferent to the outcomes. People leave their mark and, in turn, are shaped by local circumstances and cultures. They shape their worlds. The term foodshed has been coined to give an area-based grounding to the production, movement and consumption of food. Foodsheds have been described by Jack Kloppenberg as '*self-reliant, locally or regionally based food systems comprised of diversified farms using sustainable practices to supply fresher, more nutritious food stuffs to small-scale processors and consumers to whom producers are linked by the bonds of community as well as economy*'.[38]

The basic aim of regionalized foodsheds is twofold. They shorten the chain from production to consumption, thereby eliminating some of the negative transport externalities and helping to build trust between producers and consumers, and ensuring that more of the food pound gets back to farmers. They also tend to favour the production of positive environmental, social and health externalities over negative ones through the use of sustainable production systems, leading to the accumulation of renewable assets throughout the food system.

## Community-Supported Agriculture

Standing with Tom Spaulding and gazing to one horizon of this Illinois landscape, all we see is wall-to-wall yellow maize. It is a monocultural desert, except for this tiny oasis of diversity. We are at Angelic Organics, a *community-supported agriculture* (CSA) farm some two hours' drive north-west of Chicago.[39] Tom is director of the farm's Learning Centre, and he shows us a farm quite unlike anything else in the region. Angelic Organics is a 32-hectare organic farm, ten of which are used each year to grow 47 varieties of fruiting, leaf, cole, onion and root crops, and a further 12 types of herb. It is supported directly by 800 members who pay for a season's supply of vegetables in advance; each week from June to November, fresh produce is packed into boxes and delivered to Chicago, Rockford and other regional urban centres.

Unlike most of the farms in the region, this is a human-scale operation. The farm community comprises 11 staff and 3 to 5 interns, and produces 145 tonnes of vegetables per season. It is well connected to its members. It has also reached out to many other groups through its learning centre, each year providing 1000 urban young people with rural immersion experiences (few, if any, have ever been near a farm before), and horti-cultural therapy for refugees and victims of torture. A group of 150 low-income families receive free boxes of vegetables throughout the season. This is a farm connected both to nature and to its wider community, and members appreciate this fact. One member says: *'You taught me to have more respect for the work that you and other farmers do, and to appreciate and consider the connection that should exist between a healthy life and good food.'* Another writes: *'There is something to be said about being in time with the seasons. It just feels right.'* Another reflects on changed eating habits: *'We have tried so many new vegetables that I would not have bought at a store.'*

This is one of more than 1000 CSA farms across the US and Canada, the first having been established in Massachusetts in 1985. These farms directly connect with 77,000 members and bring 36 million dollars of income per year directly to the farms. The basic model is simple: con-sumers pay growers for a share of the total farm produce, and growers provide a weekly share of food of a guaranteed quality and quantity. Members typically pay US$200 to US$500 for a season's share, and would, on average, have to pay one third more for the same food at a supermarket. One study in Massachusetts indicted that a US$470 share was equivalent to US$700 worth of produce if bought conventionally.[40]

CSAs also encourage social responsibility, increase understanding of farming amongst consumers, and increase the diversity of crops grown by farmers in response to consumer demand. The central principle is that they produce what people want, instead of concentrating on crops that could give the greatest returns. In addition to receiving a weekly share of produce, CSA members often take part in life on the farm through workdays. Many CSA farms give out newsletters with the weekly food share, so that members stay in touch and know what crops are expected.

Sixty per cent of CSA farmers say that the most successful aspect of their operations is the strengthened bond with food consumers. Most offer boxes with 8 to 12 different vegetables, fruits and herbs per week; some link up with other CSAs to keep up diversity; and others offer value-added products, such as cheese, honey and bread.

In the UK, box schemes outnumber CSAs. These schemes began in the early 1990s, and now 20 large schemes and another 280 small ones are supplying several tens of thousand households weekly.[41] They ensure that good-quality produce reaches customers because food is fresh and picked the same day in the smaller schemes. Farmers contract to supply the basics, such as potatoes, carrots, onions, and one green vegetable, and add other produce depending upon the season. Over time, box schemes also increase on-farm biodiversity. In response to consumer demand, many farmers have increased the diversity of crops grown to 20–50 varieties. Prices are comparable to those in supermarkets for conventional vegetables, so consumers do not end up paying premiums. A central rationale for both CSAs and box schemes is that they emphasize that payment is not just for the food, but for support of the farm as a whole. It is the linkage between farmer and consumer that guarantees the quality of the food. This encourages social responsibility, increases the understanding of farming issues amongst consumers, and results in greater diversity in the farmed landscape. These schemes have brought back trust, human scale and a local identity to the food we eat. They also employ more people per hectare, and provide livelihoods for farm families on a much smaller area than in conventional farming.

## The Value of Farmers' Groups

Another way in which farmers can create new value in agricultural systems is to work together in groups. For as long as people have engaged in agri-culture, farming has been at least a partially collective business. Farmers have worked together on a host of activities that would be too costly, or even impossible, if performed alone. Such connections also make it easier for individuals to cross a new frontier together. There is so much that can be done with sustainable agriculture; yet, it is somehow so difficult to bring about. When there is cooperation and trust, then it is possible for new learning mechanisms to be established. Self-learning is vital for agricultural sustainability. By experimenting, farmers can increase their own awareness of what does and does not work; and if many do this together, then they rapidly multiply their learning potential.

At a meeting overlooking the sand dunes and boardwalks of Georgia's Keys, a group of North Carolina peanut growers tell their own story of change. These self-confessed former industrialized farmers had come up hard against an economic barrier. Peanuts are important in North Carolina: 2300 farmers produce 170,000 tonnes per year, the fourth largest amount by any state in the US. Since the 1930s, the Federal Peanut Programme maintained a steady and predictable price, with prices elevating whenever costs increased. But during the mid 1990s, the programme was radically changed. Prices were cut and quota carry-over eliminated, resulting in dramatic falls in farmer income.

Out of the crisis, however, emerge our heroes. With the help of Scott Marlow and colleagues at Rural Advancement Foundation International, a group of 62 farmers began reinventing both local farming and social relations. Over a period of four years, these farmers reduced pesticide use by a remarkable 87 per cent, saving themselves US$40–$50 per hectare in costs without any yield penalty. On more than 3000 hectares, they had managed to cut pesticide use by 48,000 kilogrammes. The change in attitudes and values has been rapid. A major pest of peanuts is thrips; yet most leaf damage has no yield effect, even though the crop looks damaged. By conducting their own research, farmers came to realize that they did not need to spray: *'We were farming for looks,'* says Rusty Harrell. Michael Taylor adds: *'The peanuts don't look good – but the yields increased.'*

The key to success was scientific experimentation by farmers and peer-based learning. Farmers set the agenda for field trials of alternative practices, watch for unexpected results and are encouraged to be careful about drawing conclusions. Working together, sharing experiences and developing new relationships of trust are central components of the process. *'We got together over food, and found we had a wide range of problems, and were all searching for new ways,'* says Rusty. *'We go around and look at other people's crops.'* Farmers in the group say that this has helped to bring the community together. Importantly, there are no final solutions, as sustainable agriculture needs continuous experimentation and improvement. Tom Clements says, *'This has affected our lifestyles. I'm still working on it – you have to farm true every day. Our quality of life has improved.'* The field trials gave farmers the confidence to try something new, and the trust and sharing helps them to take large steps into the unknown. As a result, incomes go up, and the environment benefits, too.

Similar changes have been provoked across the US by the government's Sustainable Agriculture Research and Extension Programme, which has supported transitions towards sustainability in a wide range of contexts. One example is work by the Kansas Rural Centre, which supports family farming and the grassroots involvement of local people in farming and countryside matters. Their Heartland Sustainable Agriculture Network brings farmers together to enhance experimentation, exchange and education. The network organizes farmers in small clusters to work together on issues that are important to them. These include Covered Acres (farmers in central Kansas experimenting with legume cover crops); Smoking Hills (farmers working on grazing management in Saline County); Resourceful Farmers (crop, livestock and dairy farmers in south-central Kansas who give on-farm demonstrations of rotational grazing and clean-water practices); and Quality Wheat (organic farmers in west Kansas seeking to improve soil fertility and increase the protein content of wheat). The network is a clearing house for ideas on sustainable agriculture, helps to build support for new ideas, nurtures leadership, creates confidence amongst farmers to try something new, and works with conventional agricultural institutions to build support for rural regeneration through sustainable agriculture.[42]

## Farmers' Markets

Farmers' markets are another simple idea, already spreading like wildfire through farm communities in both North America and the UK. Sell your produce directly to a consumer, and you get 80–90 per cent of the food pound instead of the paltry 8–10 per cent through normal marketing mechanisms. Some farmers, of course, already do this through farm shops and pick-your-own enterprises, of which there are 1500–2000 in the UK. Others are beginning to make use of direct sales by mail and via the Internet. But the best option for many is farmers' markets, which have emerged on a huge scale in recent years in the US. There were nearly 2900 farmers' markets registered with the US Department of Agriculture (USDA) in the year 2000, up from 1700 in 1994 – though some suggest that there are as many again operating at the very local level. The annual turnover in these markets is more than US$1 billion. Again, income goes directly into the pockets of the 20,000 farmers selling their produce. The USDA estimates that 6700 of these farmers now use farmers' markets as their sole marketing output. Each week, about 1 million customers visit farmers' markets, nine-tenths of whom live within 11 kilometres of the market.[43]

The benefits that these farmers' markets bring are substantial, improving access to local food, increasing returns for farmers, and contributing to community life and local cultures by bringing large numbers of people together on a regular basis. The contributions to local economies are substantial. One farmers' market in Wisconsin contributes US$5 million to the local economy each year; another in New Mexico brings an added US$700,000 to the local farmers' incomes. These farmers' markets also play a particularly important role in increasing the access for poorer families to good-quality food. Inner-city consumers typically pay one third more for their food than suburban ones, and these markets allow them ready access to wholesome and cheap food. The effect on diets can be important, too. A mid 1990s survey of New Jersey customers found that they increased their consumption of fruit and vegetables over a five-year period.

In the UK, farmers' markets have become very popular in the past four to five years. By early 2001, there were 200 established markets trading on some 3000 market days per year. In all, it is estimated that the 5 million customers at these markets each spent UK£10–£15 in 2000, thereby putting UK£50–£78 million directly into the pockets of farmers. Importantly, too, these markets are a direct connection between producers and consumers. Norman McGeoch, farmer and coordinator of farmers' markets in the eastern region, says: '*I know exactly if something is not right with my food – my customers tell me.*'[44] This may seem obvious for a business, and yet it is radical for many farmers. These farmers' markets, though, are unlikely to cause a major change in the way that most farmers market their produce. They are no answer for bulk commodities, nor will they substitute for contract sales to manufacturers and retailers. However, they do point to a vitally important principle. Where there are direct links between producers and consumers, then farmers are better able to respond to the concerns of consumers; consumers, in turn, better understand the challenges and vagaries of food production.

## Regionalized and Slow Food Systems

During recent years, some national policies have sought to link agriculture with more environmentally sensitive management. But these policies are still highly fragmented. A policy framework that integrates support for farming together with rural development and the environment could create new jobs, protect and improve natural resources, and support rural communities. Such reforms should also be supplemented with clear policy direction on regionalized food systems.[45] In North America, such integration has found meaning in localized food systems. This has received prominence owing to the effectiveness of the Community Food Security Coalition, a diverse network of anti-hunger, sustainable agriculture, environmental, community development and other food-related organiz-ations which persuaded politicians to incorporate community *food security* into the 1996 US Farm Bill. As a result, local food policy councils and systems have become increasingly effective, most notably in Hartford, Connecticut; Knoxville, Tennessee; St Paul, Minnesota; and Austin, Texas. Bringing together different *stakeholders* with common concerns and interests in a place works for local people, works for communities, works for farmers, and can benefit the natural environment.[46]

In Connecticut, the Hartford Food System (HFS) was set up by Mark Winne in order to address severe poverty and food insecurity. Some four in ten children live in poverty, and 80 per cent are eligible for free or reduced-price school meals. In low-income neighbourhoods, 25–40 per cent of residents experience hunger. The HFS promotes better food education and collective food consumption in schools. Over a period of three years, there has been a 35 per cent increase in the number of children eating breakfast at school, and a farm-to-school programme provides schools with fresh fruit and vegetables for their cafeterias. The HFS promotes urban agriculture and farmers' markets, and has initiated a coupon programme, with low-income families receiving US$10 coupons to spend at farmers' markets. As a result, four-fifths of recipients of the coupon report eating more fruit and vegetables. Similar innovations have occurred in Toronto, where a Food Policy Council has brought together an extended network of organizations concerned with food security, sustainable agriculture, public health and community development. The result has been increased fruit and vegetable consumption amongst residents; more local sourcing of foods (only one quarter of food in the social security food banks, which 150,000 people use, was sourced from Ontario farmers in 1990); and a positive effect on school children (schools with the Field to Table scheme have better attendance, less tardiness and better socialization in classrooms).[47]

The best example from Europe is the recent emergence of the Slow Food movement from Italy. This arose out of local concerns over the fast food sector's increasing homogenization and lack of responsibility towards local distinctiveness. It was founded by journalist Carlo Petrini in the mid 1980s, and now has 70,000 members in 45 countries who seek to protect local production from being driven into extinction by global brands. The idea of slow food gave rise in 1999 to the Slow City movement, which began in the four cities of Orvieto in Umbria, Greve in Tuscany, Bra in Piedmont, and Positano on the Amalfi coast. The idea of slow and distinctive food, resonant of place and people, has been taken up by local authorities, with commitments to increase pedestrian zones, reduce traffic, encourage restaurants to offer local products, directly support local farmers, increase green spaces in cities, and conserve local aesthetic traditions. Slow food and cities have given regionalized food systems and policies a name and a vision. Slow cities are also known as the *Citta del Buon Vivere* – it is, after all, about creating a good life.

Another effort to connect up food systems on a large scale comes from, perhaps, a surprising quarter. Unilever, one of the largest food businesses in the world, is developing policies and processes that will eventually allow it to source all primary agricultural produce from sustainable systems. They are assessing sustainability according to a range of tough biological, economic and social criteria, and are seeking to set standards to promote transitions for a range of produce, including peas, spinach, tea and oil palm. The central challenge, though, even for such a large operator, is to change practices throughout a whole sector. Where produce is derived from farms that have a direct relationship with a processing business, or even from its own farms, then it is relatively easy to set out new practices. But where a manufacturer buys a great deal on the open market, where it is impossible to trace products back to the farm, then the only option is to change a whole system. This is not easy and, inevitably, means moving from a stance of enlightened self-interest to one addressing wider concerns and the interests of a large number of *stakeholders*. There is, thus, an important role for small and large businesses in sustainable foodsheds.

These North American and European initiatives are good examples of the benefits of integration, and represent policy and institutional responses that can be taken, whatever the national and international policy context. There are many promising signs of progress towards sustainability in industrialized systems of agriculture. There are, equally, large forces aligned against these that are determined to capture the value of the commons before anyone notices. Perhaps it is all too late. Yet, if some of these principles are more widely adopted, then we may well see a revolution occur in industrialized farming and food systems. The principles are simple. Adopt sustainable methods of food production. Organize farmers into groups so that they can increase their marketing and purchasing power, and share experiences and knowledge on the new path towards sustainable agriculture. Organize consumers into groups, so that they can exercise greater purchasing power. Make direct links between producers and consumers so that the physical length of the food chain is shortened, consumers are sure of the quality of the food they are buying and the health of the system that produced it, and producers receive a greater proportion of the food pound.

# Concluding Comments

There appears to have been great success in industrialized food systems; yet, the seeds of destruction are now present for all to see. Alongside the disappearance of biodiversity goes the family farm, with its cultural relevance and place location. As the focus on agriculture as a commodity has grown, so farmers have come to receive a smaller proportion of what consumers spend on food. Systems of sustainable agriculture offer farmers new ways in which to reduce direct costs and dependencies on externally derived goods and services, combined with better direct connections to consumers. The concepts of bioregions and foodsheds are centred on such connectivity, and a variety of mechanisms have emerged to illustrate what can be achieved when we redesign whole systems. These include community-supported agriculture, farmers' groups, farmers' markets and slow food systems.

## Chapter 6

# *The Genetics Controversy*

## What Is Genetic Modification?

It is impossible to write of the potential for agricultural transformation without addressing the controversy surrounding *biotechnology* and genetic modification. The challenge facing us is huge, and we will need to make the very best of our collective ingenuity and willpower. This ought to mean simply making the best use of any available technology, regardless of its provenance. Who produces the technologies, how they can be made available to the poor, and whether they have an adverse environmental effect, are important questions that will tell us whether new ideas might make a real difference to the sustainability and success of agriculture. In the 21st century, we have clearly entered an information age, and information in agricultural systems is important from the smallest scale on genetics to higher scales on ecological interactions within whole ecosystems. Biotechnology represents one set of technologies that could lead to benefit. Like all new ideas, though, it requires balance and case-by-case analysis because we do not yet know all of the risks and benefits.

So, what is biotechnology? It involves making molecular changes to living or almost living things. It has a long history, dating back 4000 years to the invention of fermentation, bread-making, brewing and cheese-making by Egyptians and Sumerians, grafting techniques developed by the Greeks, and many years of selective breeding by farmers. Modern biotechnology and genetic modification are, by contrast, the terms given to the transfer of DNA from one organism to another, thereby allowing the recipient to express traits or characteristics normally associated just with the donor.[1] As these transfers or mixes do not occur in nature, the scope for genetic modification is greater than in conventional animal or plant breeding, even though advanced breeding already involves types of genetic manipulation, including clonal propagation, embryo transfer, embryo rescue and mutant selection.

The process of genetic modification involves, firstly, identifying and isolating the novel gene, called the *transgene*, as a section of DNA. This transgene codes for the production of a protein, usually an enzyme, that catalyses a novel biochemical reaction or pathway in the host plant or animal. This is then linked to a suitable promoter – another DNA sequence that regulates the expression of the gene. This construct of transgene plus promoter has to be introduced into the target organism's own chromosome. Two methods are currently available: the use of the bacterium *Agrobacterium* and the gene gun. *Agrobacterium* naturally transfers DNA to its host plant, causing diseases or the formation of galls. But for the purposes of genetic modification, its plant-gall inducing capability has been removed, and it works as a vector to transfer DNA. Initially, this method only worked for broadleaf plants, but has now been developed for transforming cereals. The gene gun, by contrast, fires microparticles of gold or tungsten coated with the transgene constructs into the target cells.

Neither process, though, is predictable since incorporation of the transgenes into the host DNA is largely random. Location in the genome is vital, as only some of the individual organisms will express the desired characteristics. Once these have been identified, they are grown and bred conventionally. This process of identification requires the use of a selective marker – some way to distinguish at cellular level the cells that contain the transgene and those that do not. The construct thus contains a third element – a marker gene. The easiest markers confer resistance to anti-biotics or herbicides, so that non-genetically modified cells can easily be filtered out. However, antibiotic markers are a cause for concern, given their overuse in farming and medicine, combined with the growth of antibiotic resistance. A range of alternatives is therefore being actively developed, such as staining, fluorescence and reporter genes.

The biotechnology industry has grown rapidly in both the food and medical sectors in recent years. The first genetically modified (GM) products eaten by humans were cheese and tomatoes. GM bacteria were first used in the early 1990s to produce chymosin, an alternative enzyme to calf rennet, for vegetarian cheese (the GM bacteria are not eaten). Then, in 1995, the first year to see commercial cultivation of GM crops anywhere in the world, tomatoes with their softening gene inactivated, allowing them to ripen until they reach full flavour and colour without rotting, were marketed as tomato paste. Since then, the greatest commercial growth has been in crops containing one of two traits. These comprise, firstly, herbicide tolerance, introduced in soya, oilseed rape, cotton, maize and sugar beet, which allows the application of broad-spectrum herbicides to the crop, thereby killing all of the weeds without damaging the crop. The second trait is insect-resistance through expression of a gene from the bacterium *Bacillus thuringiensis (B.t.)*, mainly in maize and cotton, which means that the *B.t.* insecticidal toxin is expressed by all cells of the plant, thereby killing susceptible pests and so reducing the need to apply some conventional insecticides. By 2001, there were 50 million *hectares* worldwide, about three-quarters of which were in the US, and most of the rest in Canada, Argentina and China. In Europe, small amounts were commercially grown in France, Spain, Portugal and the Ukraine.[2]

# New Developments in Medicine and Agriculture

Coinciding with their development in agriculture, these genetic-modification techniques are also being used in medicine for the study of genes and their function, and the replacement of genes that cause disease. Gene therapy will provide opportunities for curing some hitherto untreatable diseases. One is cystic fibrosis. This affects some 50,000 people worldwide, and damages their respiratory and intestinal tracts. An inability to clear mucus from these organs leads to intestinal blockage and recurrent chest infections, eventually causing respiratory failure. Once the mutated gene responsible for cystic fibrosis had been identified, this opened the way to replace the mutant gene with a normal copy, introduced by a vector into the lungs. Although much research remains to be done, it is now likely that a complete cure could soon be developed. Other candidates for gene therapy include muscular dystrophy and heart disease.[3]

Molecule 'pharming' is the term used to describe the use of animals and plants to make pharmaceutical products for medical applications. In principle, virtually any molecule produced by the human body can be made in a genetically modified animal or plant. Human proteins can be grown and harvested like any other crop. The current technology involves fermentation with micro-organisms in a bioreactor; but 'pharming' with genetic modification is likely to be more controllable and efficient. Sheep and pigs have already been modified to produce human proteins in their milk, such as insulin, interferon, and the human blood-clotting protein factor-eight, which is vital for haemophilia sufferers because it is free from human viruses. Rice has also been engineered in California to produce alpha-antitrypsin, a human protein used to treat liver disease and haemorrhages. The transgenic rice is grown normally, harvested and allowed to malt. Normally, it produces an enzyme that turns starch into sugars, but it has been modified to produce the human protein rather than the enzyme. In the UK, alpha-antitrypsin is produced by transgenic sheep, and Dolly, the first cloned sheep, was created in order to allow multiple copies to be made of animals without diluting valuable genetic traits through conventional breeding.

During the late 1990s, genetically modified organisms were producing one quarter of all insulin, growth hormone, hepatitis-B vaccine, and monoclonal antibodies needed for cancer treatment. Today, other medical applications under development include gene treatments for multiple sclerosis sufferers, and blood vessel drenches with DNA to encourage human hearts to grow their own bypasses. All of these medical applications are likely to bring substantial public and consumer benefit, though none is, of course, entirely without risk.

Most of the agricultural applications of genetic modification to date represent changes to 'input-traits', or genes that control specific plant functions, such as herbicide tolerance or insect resistance. Many new developments will be in so-called 'output-traits', in which farm products could be redesigned to meet specific farmers' circumstances or customers' needs, though whether these represent desirable or low-risk opportunities is another matter. Plants and animals could be modified to deliver a wide range of drugs, plastics, oils, human proteins and other products of social value. In future, some farms (or perhaps 'pharms') will produce these products rather than just food or fibre. Plants could be engineered with drought, salt, thermo, frost and aluminium tolerance, so that degraded and hostile environments could be opened up for food production. Some 10 per cent of the irrigated land in the world (27 million hectares) suffers from extreme salinity, and a further 20 per cent has symptoms of salt damage. Could these lands be turned into productive ones? Work is also underway to incorporate genes from a cold-dwelling fish into sugar beet, tomatoes, strawberries and potatoes, thereby conferring the host plants with a new mechanism for frost tolerance.

Maize, soya beans, oilseed rape and other oil crops could be modified to alter their saturated fat content. A potato with a higher starch content would absorb less oil during frying, providing an alternate method of producing lower fat products such as chips and crisps. Some fruits and vegetables will be adapted to contain higher levels of vitamins C and E. Blue cotton has been engineered through the transfer of a gene from an unnamed blue flower, potentially eliminating the need for dye. In time, fruits and vegetables could be produced in different colours, though whether we would want this is another matter. Another possibility is that fruits and vegetables will be engineered with genes from pathogenic viruses and bacteria so that, when consumed, they will encourage the production of antibodies without the recipient having been exposed to the harmful organism. Vaccine potatoes that confer resistance to *E.coli*-caused diarrhoea have already been tested, and banana vaccines are under development. In time, oral vaccines in fruits could replace conventional vaccines. A far more difficult problem is the genetic engineering of nitrogen fixation, with the distant possibility that cereals could fix their own nitrogen with the help of rhizobia associated with their roots, thereby reducing or even eliminating the need for inorganic nitrogen fertilizers. But the process would have to involve engineering symbiotic bacteria, and then persuading them to create stable and heritable relationships with the cereal.

The many potential agricultural and medical applications of genetic modification do, however, raise fundamentally important ethical issues. Xenotransplantation, involving the transplant of animal organs into humans, could meet the high demand for organs for transplantation. In the UK, there are more than 5000 people on the waiting list for organ transplants. Genetic modification offers the opportunity to create new organs in modified pigs. But, to date, the risks of encouraging the spread of retroviruses from pigs to humans outweigh potential benefits. Genetic modification also opens the way to the body-part shop; companies in the US are already working on creating skin, veins, bone, liver, cartilage and breast tissue. It also raises the spectre of pollution-tolerant humans – individuals with genes that confer tolerance to poisonous chemicals who would be able to, or perhaps made to, work in places where such pollution is widespread. Human reproductive *cloning*, once thought to be far away in the realms of science fiction, is likely one day to become fact. New information on an individual's genes could also be misused, with the possible emergence of a new genetic 'underclass' unable to get life insurance.

## Divided Camps and Different Technologies

Only a few years after the development of the first genetically modified crops for agriculture, opinions on benefits and risk are sharply divided. Some argue that genetically modified organisms are safe and essential for world progress; others state that they are not needed and hold too many risks. The first group believes that media manipulation and scare-mongering are limiting useful technologies; the second that scientists, private companies and regulators are understating hazards for the sake of economic returns.[4]

Neither view is entirely correct for one simple reason. Genetically modified organisms are not a single, simple technology. Each product brings different potential benefits for different *stakeholders*; each poses different environmental and health risks. It is, therefore, useful to distinguish between different generations of genetically modified techno-logies. The first-generation technologies came into commercial use in the late 1990s and have tended not to bring distinct consumer benefits; this is one reason why there is so much current public opposition. The realization of promised benefits to farmers and the environment has only been patchy. First-generation technologies include herbicide tolerance, insect resistance, long-life tomatoes, bacteria in containment for the production of cheese and washing-powder enzymes, and pre-coloured flowers and cotton, such as black carnations and blue cotton.

The second-generation technologies comprise those already developed and tested, but not yet commercially released, either because of uncert-ainties over the stability of the technology itself, or over concerns for potential environmental risks. Some of these applications are likely to bring more public and consumer benefits, and include a range of medical applications. These include viral resistance in rice, cassava, papaya, sweet potatoes and pepper; nematode resistance in various cereal and other crops, such as banana and potato; frost tolerance in strawberry; *B.t.* clover; trees with reduced lignin; vitamin-A rice; and 'pharming' with crops and animals for pharmaceuticals.

The third-generation technologies are those that are still far from market, but generally require the better understanding of whole gene complexes that control such traits as drought or salt tolerance, and nitrogen fixation. These are likely to bring more explicit consumer benefits than the first generation. These include stress tolerance in cereals, such as thermo, salt and heavy-metal tolerance; drought resistance; physiological modifications of crops and trees to increase efficiency of resource use (nutrients, water, light) or delaying of ageing in leaves; neutraceuticals (crops boosted with vitamins/minerals); vaccine crops (such as banana and potato); designer crops modified to produce oils or plastics; the development of new markers to replace antibiotics; and legumes with increased tannins for bloat control in cattle.

The first-generation technologies have tended to provide substantial private benefits for the companies producing them. Herbicide-tolerant soya, for example, locks farmers into buying the herbicide produced by the company who markets the genetically modified seed. Many of the later-generation genetically modified organisms are, by contrast, more *multifunctional* and *public-good* oriented; though clearly none is without risk. Modifications of crops with low value in rotations, such as legumes and oats, will make them more attractive to farmers because of high protein and energy content. Others will be more efficient in nitrogen use, thereby reducing nitrate leaching; or modifications of rhizobia could improve the nitrogen-fixing capacity of a wide range of crops. Both options would reduce the need to use nitrogen fertilizers.

A breakthrough in plant breeding would occur with the transfer of apomictic traits into cereals – the production of exact clones of the mother plant through asexual reproduction. Research in Mexico by the International Centre for Maize and Wheat Research is seeking to transfer *apomixis*, a trait involving several genes, from a grassy relative of maize, *Tripsacum dactyloides,* to maize itself. This would turn currently higher yielding but infertile hybrid seeds into fertile ones, allowing farmers to save the seed for subsequent seasons. This could boost the yields of poorer and remote farms, provided a means could be found to get the seeds to farmers when needed. This technology contrasts with terminator techno-logy – an application of genetic modification more for public benefit. There are already concerns, however, that many of the methodologies and products in this process of GM apomixis transfer are being patented by companies, and therefore will not become available to poorer farmers. In 1998, the Bellagio Apomixis Declaration was formulated, with signatories sharing a concern that the *'current trend towards consolidation of plant biotechnology ownership in a few hands may severely restrict affordable apomixis technology, especially for resource poor farmers'.* Clearly, property relations are crucial in deciding whether such developments will confer public benefits.[5]

# The Environmental and Health Risks of Genetically Modified Crops

Agricultural genetically modified organisms pose a range of potential environmental and health risks.[6] These include five types of environmental risk and two risks for human health. The degree to which each of these poses an actual risk is a combination of both a hazard and exposure, since not all hazards constitute a risk in practice. Thus, the risks and potential benefits are different for every application of genetic modification. Each class of risk is analysed below in light of recent independent scientific knowledge, drawing particularly upon analyses from the field.[7]

## Gene flow

The first potential environmental risk is gene flow, where transgenes could transfer from a genetically modified organism to wild relatives and/or bacteria in soil or human guts. Gene flow is a natural phenomenon, with many species of plants crossing with related species. As a result, the question of novel risk rests on whether the transgenes could lead to the transfer of undesirable traits, and the emergence of permanently transformed populations. As these transfers have not occurred in nature, it is impossible to predict the effects with confidence.[8] The main concerns lie in pollen transfer. However, it is important to note that pollination is not the same as gene flow; although pollen can travel many kilometres, only rarely will it result in a fertilization event.[9] Furthermore, many genetically modified lines are male sterile; so, even though pollen transfer may occur, pollination cannot. A further concern is the potential for uptake of transgenic DNA by soil bacteria, which is referred to as horizontal gene flow.[10] The important question is not so much whether gene flow occurs; rather, to what extent might transgenes affect native plant ecology? As Brian Johnson of English Nature put it: *'To add genes from other plants unwittingly and randomly to native gene pools may result in phenotypic effects which could change the way entire genomes relate to their physical and biotic environments.'*[11] Thus, the transfer of transgenes that are designed to prevent germination would lower fitness of new crop-native hybrids, whereas resistance to insects, fungi and viruses could substantially increase fitness. This could lead to the emergence of weeds with multiply stacked genes for herbicide tolerance.

## Emergence of new forms of resistance and secondary pest and weed problems

The second environmental risk centres on the potential for the emergence of new forms of resistance and/or secondary pests and weeds. Resistance had already emerged on a very large scale in modern agriculture before the advent of genetically modified organisms. There are now 500 species of insect, mite or tick that are resistant to one or more compounds, together with more than 400 herbicide-resistant weed biotypes, and 150 resistant fungi and bacteria.[12] Evolution of resistance can occur in the context of genetically modified crops that express an insecticidal product (eg *B.t.*), leading to insect resistance, or through overuse of herbicides on genetically modified crops, causing weed resistance. At first, the potential problem of insect resistance went unrecognized. Now, though, there are mandatory rules in the US to reduce the selection pressure on pests through integrated resistance management for *B.t.* genetically modified crops. These mandate that a proportion of the cropped area must be devoted to refuges of non-genetically modified crops, that rotations must be used, and that *B.t.* maize should not be grown where pest pressure is low. The guidance indicates that 20 per cent of farmland must be devoted to refuges within 800 metres of a *B.t.* crop field, with varying rules for refuge size depending upon the proportion of a parish under the same genetically modified crop. For *B.t.* maize grown in a cotton area, the stipulation is a 50 per cent non-*B.t.* maize refuge in order to minimize corn earworm and cotton bollworm resistance. The aim is to provide sufficient susceptible adult insects to mate with potential *B.t.*-resistant adult insects in order to dilute the frequency of resistance genes. But there is still controversy over the size, structure and deployment of non-*B.t.* crop refuges, how they should be implemented at a regional scale, and the difficulty of enforcing or encouraging farmers to adopt them.

## Recombination of viruses and bacteria to produce new pathogens

A third risk relates to the potential for viruses or bacteria to incorporate transgenes into their genomes, leading to the expression of novel and possibly undesirable traits. In addition, viral transgenes that are incorporated into the genetically modified crop could, in theory, recombine to produce viruses with high fitness. However, such recombination has not yet been shown to occur.[13] In theory, viral genes could affect humans, too, by surviving passage through the human gut and entering gut bacteria and human body cells. Once inside cells, DNA could insert itself into the genome to change the basic structure and functions. This could lead to the emergence of new diseases. However, this would necessitate the highly improbable integration of whole sequences of DNA into the human genome.

## Direct and indirect effects of novel toxins

The fourth risk centres on the potential direct and indirect effects of novel toxins expressed by genetically modified organisms. *B.t.* is expressed by all cells in a *B.t.* maize or cotton plant, and therefore could affect either beneficial organisms coming into direct contact with the plant or plant products, or, indirectly, through consumption of a herbivorous insect that has sequestered the toxin in its tissue. In laboratory conditions, several potential risks have been demonstrated, such as genetically modified potatoes that express a lectin; *B.t.* maize that affects ladybirds, lacewings and butterflies; and *B.t.* products in the soil. However, these laboratory studies do not necessarily mean that a real risk arises in the field.[14]

A good example of the difficulties encountered is represented by recent studies of the effect of pollen from genetically modified maize on monarch butterflies (*Danaus plexippus*). The larvae of monarchs were reared in laboratories on milkweed leaves dusted with *B.t.* maize pollen, and these larvae ate less, grew more slowly and had higher mortality than those reared on leaves dusted with non-genetically modified pollen. The potential threat to a nationally important species raised great concerns about genetically modified organisms in general, despite the fact that *B.t.* is already known to be toxic to Lepidoptera. However, the dose of pollen required to cause an effect in the field, the amount of pollen on milkweed leaves, the likelihood of butterflies being exposed to pollen, and the photodegradation of *B.t.* and rain-washing effects all remain unknown. For monarchs, timing is vital. In order for harm to occur, the larvae have to emerge at the same time as maize is pollinating, a narrow period of seven to ten days. However, monarch migration and *B.t.* pollen show only coincide in certain areas; pollen does not travel far (90 per cent falls in the first 5 metres); larvae on milkweed are not adversely affected by *B.t.* pollen; and most milkweed tends not to be found close to maize fields. Moreover, only one form of *B.t.* has been found to be consistently toxic to monarchs. Again, this does not mean that all potential risks from *B.t.* crops will be small, or even that all insects will not be harmed – just that a detailed understanding of the context of the cropped environment is needed before a clear judgement about risk can be made.[15]

## Changes to farm practices leading to changes in biodiversity

As a result of the incorporation of genetically modified organisms within their farm practices, farmers may also contribute directly or indirectly to biodiversity losses. The primary concern centres on the adoption of herbicide-tolerant crops that result in the increased use of broad-spectrum herbicides. Such products offer the option of a 'complete weed kill', which is good for the crops, but particularly bad for farmland plants, mammals and birds. The trend towards clean fields with no weeds, and thus no herbivorous insects or seed production (which, in turn, comprise food for birds and mammals), has been a major factor in the decline of farmland birds.[16] Once again, however, much depends upon the detailed agronomy and goals of farmers. Some genetically modified organisms could lead to greater biodiversity. Recent research shows that glyphosate-tolerant sugar beet can reduce annual input costs from UK£230 per hectare (not counting the company's technology fee), with farmers able to leave weed control until at least the four-leaf stage, thus making beet plants harder for aphids to find and encouraging beneficial predators. This precise control of weeds during the time when they pose a real threat to yields could also give the option of greater tolerance of weeds at other times, thereby leading to biodiversity benefits. At the same time, however, glufosinate ammonium-tolerant sugar beet has been shown to allow virtually complete removal of all weeds using less herbicide than a conventional crop would require. In the US, detailed studies have shown that some farmers with herbicide-tolerant soyabeans are surprisingly using two to five times more herbicide than conventional growers.[17]

## Allergenic and immune system reactions to new substances

Since transgenes result in the manufacture of new products in crops, usually proteins, a risk to humans arises if these products provoke an additional allergenic or immune response. Conventional non-genetically modified foods already contain a large number of toxic and potentially toxic products. As a result, the key question is whether a specific genetically modified organism could result in a new hazard. As 90 per cent of food allergens occur in response to proteins found in eight foods, namely peanuts, tree nuts, milk, egg, soyabean, shellfish, fish and wheat, it could be argued that as genetic modification involves transfer of a single or a few genes, so it is easier to test for allergenicity. One product, genetically modified soya with a brazil nut gene, was withdrawn from development because of potential allergenic effects.[18] The greatest controversy has surrounded the case of genetically modified potatoes containing lectin and their effect on rats. Immune response effects have been claimed, but the research has been widely criticized. If the research had, indeed, shown an effect, then this would be significant only for this particular gene and its product. Equally, though, the absence of effect does not mean that all genetically modified organisms are safe. Other potential problems might arise in potatoes with modified biochemical pathways that could inadvertently lead to increased levels of glycoalkaloids. It is also important to distinguish between the consumption of food products that potentially contain genetically modified DNA, and food products that are identical to those from conventional crops, such as refined sugar, which contains no DNA.[19]

## Antibiotic resistance marker genes

The first-generation genetically modified organisms have used antibiotic or herbicide marker genes for easy cellular selection. In theory, antibiotic-resistant marker genes from a genetically modified organism could be incorporated into bacteria in the guts of humans and livestock, rendering them resistant to the antibiotic. Although this has not yet been demonstrated empirically, antibiotic resistance is still a major cause for concern. Antibiotics and other *antimicrobials* are used in agriculture for therapeutic treatment of clinical diseases (20 per cent) and prophylactic use and growth promotion (80 per cent of total). Concern is growing that the overuse of antibiotics may render some human drugs ineffective and/or make some strains of bacteria untreatable. The World Health Organization has documented direct evidence that antimicrobial use in farm livestock has resulted in the emergence of resistant *Salmonella, Campylobacter, E.coli* types, and vancomycin-resistant Enterococci that are linked to the overuse of antibiotics both in hospitals and on farms.[20] Alternatives to antibiotic markers now exist, and many believe antibiotics should not be used in commercial genetically modified organisms.[21] The Royal Society has said: *'It is no longer acceptable to have antibiotic resistance genes present in a new genetically modified crop.'*[22] Nonetheless, it is still not clear whether antibiotic marker genes add significantly to the risk of resistance that is emerging from exposure to antibiotics used elsewhere in the food chain.

## The Contrasting Concerns of Different Stakeholders

The pace of change in developing genetic modification has provoked many debates, some specifically about the benefits and risks of genetically modified technologies. Others, though, are about important indirect effects, such as the growing centralization of world agriculture, that represent structural changes in agriculture in which genetically modified organisms are a contributor to change, but not necessarily the driving factor. These contested positions raise important questions. Will genetically modified organisms contribute to the singular promotion of technological approaches to modern agriculture, or could such technologies bring environmental benefits and promote sustainability? Are genetically modified technologies essential for feeding a hungry world, or is hunger more a result of poverty, with poor consumers and farmers unable to afford modern, expensive technologies? In addition, does genetic modification across species represent a breakdown of natural species barriers, or does the presence of common gene sequences in very different species indicate that such transfers are part of evolutionary history, and therefore of little novel concern? Are foods produced from genetically modified organisms 'substantially equivalent' to other foods, and therefore do not require labelling, or is labelling a right for consumers because it permits them to make informed choices? Will genetically modified organisms contribute to greater consolidation of corporate power in the food system; and even if they do, are such globalized operations a necessary and desirable part of economic growth?

There are no simple answers, and this has brought great confusion and a tendency for the protagonists to dismiss the concerns of environmental or consumer groups as misguided, but without realizing how complex are the concerns of people when promises are made about new technologies. Equally, those against genetically modified organisms too readily dismiss the pro-lobby as unbalanced in presentation and unable properly to assess the case-by-case risks.[23] A significant danger is that scientists, together with farmers who produce the food, will further lose the trust of citizens. Mary Shelley's Dr Frankenstein is condemned not so much for what he wanted to achieve, even though it may have been flawed, but because he failed to take responsibility for his actions.[24] The creature, popularly but incorrectly called Frankenstein, does not engage in gratuitous violence. Rather, because he is lonely, he takes revenge when the scientist, Frankenstein, refuses to create another companion for him. Lack of responsibility and trust could irreparably damage the science of genetic modification. Many food manufacturers and retailers have banned genetically modified products from their foods. Many farmers are uncertain. They would like access to technologies that may give competitors an advantage; but, equally, they would not like to lose the trust of consumers any further.

Yet, there is much that can be done to engage wider groups of stakeholders in constructive debate and discussion, and to ensure the adoption of a cautious and evidential-based stance towards new technologies. Tim

O'Riordan of the University of East Anglia has suggested some guidance for such a stance.[25] Where unambiguous scientific proof of cause and effect is not available, then people must act with a duty of care. Where the benefits of early action are judged to be greater than the likely costs of delay, it is appropriate to take a lead and thus inform why such action is being taken. Where there is the possibility of irreversible damage to natural life-support functions, precautionary action should be taken irrespective of the forgone benefits. Individuals should always listen to calls for a change of course, incorporate representatives of such calls into deliberative forums, and maintain transparency throughout. Individuals, organizations and governments should never shy away from publicity and try to suppress information, however unpalatable – in the age of the Internet, someone is bound to find out if information is being distorted or hidden. Finally, where there is public unease, it is important to act decisively in order to respond to that unease by introducing extensive discussions and deliberative processes. This is so that benefits and costs can be discussed together.

Not all agree, however, on the value of such deliberation. The US Senate Committee on Science, for example, adopted a highly combative tone when reporting on genetically modified organisms in the US. It was dismissive of 'political activists', indicating that critics of genetic modification had '*mounted a well-funded campaign*', as if it was unfair that they should also be well funded. It is unlikely that this continuing dismissal, on both sides, will lead to constructive outcomes.[26]

## Genetic Modification: Another Technological Fix or a Contributor to Sustainability?

Another area of disagreement concerns the potential for genetically modified organisms to contribute to greater sustainability in agriculture. The issue depends fundamentally upon the technologies and practices that genetically modified technology would replace. For example, a technology resulting in the reduced use of pesticides would be more sustainable than a conventional system relying on pesticides; but this reduced-use system would score less well if compared with an organic system that used no pesticides.

Many commentators have argued that genetically modified technology represents no more than a further technological fix on an intense agricultural treadmill. Modern agriculture has been highly successful at increasing food production; but it has also brought costly environmental and social consequences. Solving these problems has often meant treating the symptoms rather than the underlying problems. In this process of technological determinism, technology is seen as a 'cure-all' for problems; the tendency is to address the symptoms rather than underlying causes. Miguel Altieri of the University of California at Berkeley is worried that '*biotechnology is being pursued to patch up the problems caused by previous agrochemical technologies (pesticide resistance, pollution, soil degradation) which were promoted by the same companies now leading the bio-revolution*'.[27]

To what extent, then, are commercially cultivated genetically modified organisms currently contributing to transitions towards sustainability? It is important to note that not all commercially cultivated genetically modified organisms are alike in their outcomes, despite what some individuals say about genetically modified organisms both increasing yields and reducing agrochemical use. Unconditional claims by companies, or by industry-funded research, have fostered further questions about the efficiency of genetically modified technologies. For every company press release or aligned report that indicates substantial yield and environmental benefits, there is another report that suggests problems with the technology. It is impossible to draw any firm conclusions from either side.[28]

Well-designed and independent research takes longer to conduct and write up, and it was only after a few years of cultivation that field-based evidence appeared. Independent research from the Universities of Arkansas, Missouri, Nebraska, Ohio State, Purdue and Wisconsin conducted during 1999–2000, together with some reports from the US Department of Agriculture and the US Environmental Protection Agency, indicated a highly mixed performance in the field, including some agronomic surprises. This literature does not support the US Senate Committee on Science's broad contention that '*the current generation of pest-resistant and herbicide-tolerant agricultural plants produced by biotechnology has reduced chemical inputs and improved yields*'. In reality, there were some substantial increases in herbicide use and some falls, and there were some significant reductions in total insecticide use – although this amounted to relatively little on a per hectare basis.[29]

## Genetic Modification: Driver of Corporate Power or Friend of Farmers?

Another contested issue relates to the rapidly changing structure of world agriculture, especially the vertical integration of corporations, and the growing concentration at every stage of the food chain. There are fewer input suppliers, farms, millers, slaughterers, packing businesses, and processors than ever before. Such vertical integration is a concern to many, with the UK House of Lords stating: *There is a concern, shared by farmers, witnesses and ourselves, that the powers of a few agrochemical/seed companies are already great, and will become greater, over the process of producing (developing and growing) GM crops.*[30]

Since many genetically modified organisms are being commercially produced by large corporations, there is intense interest in how power relations and property rights will play out.[31] Important questions arise. To what extent, for example, are these private interests concerned only with their shareholders' gain, or are they willing to engage with farmers of all types, both in industrialized and developing countries? For the first generation of genetically modified crops, reduced use of insecticides, combined with increased yields, should mean greater benefits for farmers. Companies, however, charge a technology fee, on top of seed costs; to date, this appears to capture most or all of the margin in certain systems. But if the genetically modified organism fails to deliver promised benefits to farmers, then corporate–farmer relations may begin to fail. In 1998, 55 Mississippi farmers complained to their state department of agriculture and commerce's arbitration council on the grounds that their genetically modified cotton had lower yields or had completely failed. Most settled out of court; three were awarded nearly US$2 million in damages. A year later, 200 cotton farmers from Georgia, Florida and North Carolina were engaged in a legal dispute with Monsanto after crop failure of *B.t.* and herbicide-tolerant cotton.

A critical issue relates to who gets (or owns) the benefits of the new technology. Patent law is vital because it treats genes and genetic engineering in the same way as any other invention. To be patented in Europe, as covered by the European Convention, an invention must be 'new', 'not obvious', 'capable of industrial application' and 'patentable subject matter'. An invention must add to the current state of knowledge. A new method of isolating a gene qualifies, as does an isolated gene with a new activity; but a gene in a human body does not qualify. It is possible, however, to patent an artificially synthesised gene or the replication of the genetic information contained in the gene. The international Convention on Biological Diversity (CBD) is important for property rights. It came into force in December 1993, and has three aims – namely, the conservation of biological diversity, the sustainable use of its components, and the fair and equitable sharing of the benefits arising from genetic resources. However, it remains difficult to allocate 'ownership' when genes interact in highly complex ways to express characteristics. The conventional wheat variety Veery, for example, was the product of more than 3000 different crosses involving parents from 26 countries. Under the CBD, the country of origin and the legal owner of plant genetic resources are legally defined as the first to file a claim on ownership; but it is very difficult to attribute clear ownership when a variety is derived from so many sources.[32]

There are signs, however, that some corporations are developing new benefit-sharing mechanisms. A ground-breaking arrangement between AstraZeneca, now Syngenta, and the inventors of vitamin-A rice, also called golden rice, will permit farmers in developing countries to earn up to US$10,000 without paying royalties.[33] The deal permits the company to commercialize the rice, while effectively providing it free to small farmers. There remain, however, many controversies over so-called 'golden' rice, including the cultural resistance to eating orange-coloured rice, the need for adequate irrigation, and whether vitamin-A deficiency could better be addressed through diversified diets. Another example is the Positech selection technique, an alternative to antibiotic resistant markers. Developed by Novartis, now also Syngenta, at a cost of US$10 million, the company has said it will market Positech under a two-tier pricing system, with commercial uses incurring royalties, while those developing technologies for subsistence farmers will be granted free access. But a drawback of this means that public-funded researchers are often unable to use such technologies, owing to their high price.

## Genetic Modification: Feeder of the World or Eliminator of Alternatives?

A further debate centres on whether genetically modified crops could help to feed the world. Some emphatically say yes, often raising the spectre of famine as a way to gain greater support for genetic modification as a whole.[34] But genetically modified technologies can only help to feed the world if attention is paid to the processes of technology development, benefit-sharing and, more especially, to alternative or low-cost methods of production. Most commentators agree that food production will have to increase, and that this will have to come from existing farmland. But past approaches to modern agricultural development have not been successful in all parts of the world. In most cases, people are hungry because they are poor. They simply do not have the money to buy either the food they need or the modern technologies that could increase their yields. What they need are readily available and cheap means to improve their farm productivity. As a result, a cereal crop engineered to have bacteria on the roots to fix free nitrogen from the air, or another with the apomixies trait, would be a great benefit for poor farmers. However, unless this technology is cheap, it is unlikely to be accessible to the very people who need it most.

As indicated elsewhere in this book, sustainable agriculture is now an increasingly viable option for many farmers in developing and industrialized countries alike. But where there are no alternatives to specific problems, then genetic modification could bring forth novel and effective options. If research is conducted by public-interest bodies, such as universities, non-governmental organizations and governments themselves, whose concern it is to produce *public goods*, then biotechnology could result in the spread of technologies that have immense benefits. Research that is likely to bring new options for farmers already includes studies on virus-resistant cassava, potatoes, sweet potatoes, rice and maize, nematode-resistant bananas, thermo-tolerant and drought-tolerant pearl millet, *Striga*-resistant maize, and pest-resistant wheat.[35]

One good example is rice yellow-mottle virus, which is a major factor in limiting African rice production, often reducing yields by 50–95 per cent.[36] It has not been possible to introduce resistance into local varieties through conventional breeding; but genetic modification has led to the development of novel resistant varieties. These have been tested in five countries, resulting in complete resistance to the virus. Another example is tolerance to salinity, which affects 340 million hectares of land worldwide. Some plants are known to produce and accumulate osmo-protectant solutes, such as glycinebetamine, mannitol, trehalose and proline. These non-toxic solutes can accumulate to osmotically significant levels in order to protect against damage from high salt concentrations in the soil. Introduction of single genes has led to modest accumulations of solutes, However, to be successful, multiple-genes coding for entirely new metabolic pathways will be needed.

Further applications could improve yields in developing countries if they remove or tolerate a stress, such as rice that tolerates prolonged submergence, and if they allow cultivation of problem soils, such as those affected by aluminium toxicity.[37] Nonetheless, new threats to the livelihoods of developing country farmers may yet arise. Transgenic tropical crops, such as sugar cane, oil palm, coconut, vanilla and cocoa, could be grown in temperate countries with appropriate genetic modification. Other crops may be engineered to replace tropical products. Oilseed rape, for example, could be engineered to produce lauric acid for soap-making, thereby threatening producers of oil palm in Malaysia and Ghana.

# Further Policy Directions

Genetically modified organisms are not a single, homogeneous technology. Each application brings different potential benefits and risks for different stakeholders. Regulators, therefore, face special challenges in the face of rapidly developing technical applications. In the European Union, releases of genetically modified organisms were regulated under Directive 90/220 for a decade. Following protracted negotiations, this has now been revised, harmonized and tightened, and signed into effect in early 2001. The new directive sets out provisions for the scientific assessment of the risks of experimental and commercial releases of genetically modified organisms into the environment, and establishes protocols for post-release monitoring.

To date, the general approach to risk assessment in agriculture, as a whole, has been to establish rigorous procedures prior to release, and then to assume that farmers engage in 'good agricultural practice'. The novel nature of emerging policies centres on a fundamental shift in risk assessment to a need to understand the effects of technologies in the field and on the farm. Much of the harm to the environment arises when technologies, whether pesticides, fertilizers or machinery, are not used in accordance with regulators' criteria. The assessment of genetically modified organisms will, however, now contain new requirements to assess the effects of diverse farm practices on the genetically modified organisms themselves, and to determine how this interaction will affect desirable environmental outcomes, such as the integrity of local biodiversity. Such new risk assessments could have a positive side effect by increasing our understanding of agricultural–environment interactions in agricultural systems at large.

However, these standards for regulation are not yet widespread. The challenge that developing countries face is to find ways of increasing regulatory and scientific capacity in order to assess the effects of modern agricultural technology on their environments. The Convention on Biological Diversity establishes a broad framework for assessing effects. Efforts are underway to see the January 2000 agreement on adopting the precautionary principle as the basis for an international biosafety protocol, and ratified by 130 countries, signed and put into practice.[38] The centre piece could be an 'advance informed agreement' procedure to be followed before transboundary transfer of genetically modified organisms, although a bloc of agricultural exporting nations still argue that agricultural commodities should be excluded from this procedure. Whether such international agreements can be signed or not, there is still a high priority on findings ways to help build domestic scientific and legal expertise within countries in order to establish comprehensive biosafety protocols for genetically modified organisms.[39] Such policy frameworks will need to protect intellectual property rights, to protect against environmental and health risks, and to regulate the private sector if developing countries are to benefit significantly from genetically modified technologies. It seems likely that biotechnology will make some contributions to the sustainability of agricultural systems. But for the poorest farmers, communities and countries, biotechnology is unlikely to make a very significant contribution for some years. As indicated earlier, a significant priority is the maximization of benefits from agroecological approaches that rely on high ecological literacy and good social relations.

## Concluding Comments

In this chapter, I have addressed the genetics controversy in agricultural systems. It is impossible to think about agricultural transformation without assessing these technologies, and without appraising who is producing them and what they could bring in the form of both benefits and costs. There are many applications of biotechnology, and there are likely to be several distinct generations of released technologies. It would be wrong, therefore, to generalize about genetic modification – each application needs to be addressed on a case-by-case basis. We need to ask questions about who produces each technology and why; whether it can benefit the poorest and, if so, how will they access it; and whether it will have adverse or positive environmental and health side effects. It is likely that biotechnology will make some contributions to agricultural sustainability; but developing the research systems, institutions, and policies to make them pro-poor will be more difficult.

## Chapter 7

# *Ecological Literacy*

## Knowledges of Nature

Despite great technological advances in agriculture, the value of the knowledges and practices of local communities is only slowly being acknowledged. We often use the word traditional, yet it remains a problematic term. To many, it implies a backward step – knowledge wrapped up in superstition or quaint old ways – and there is no place for this in our modern world. Traditional, though, is best thought of as not a body of knowledge itself, but the process of knowing. If our lives involve continuous writing and rewriting of our own stories, through adjusting behaviour, incorporating new understanding into our cultures, and shaping and being shaped by local nature, then knowledges are also undergoing continuous revision. Darrell Posey, anthropologist and protector of the rights of the excluded, quoted the Four Directions Council of Canada to produce a compelling definition:

*What is 'traditional' about traditional knowledge is not its antiquity, but the way it is acquired and used. In other words, the social process of learning and sharing knowledge. . . Much of this knowledge is quite new, but it has a social meaning, and legal character, entirely unlike other knowledge.*[1]

An acquisition process such as this inevitably leads to greater diversity of cultures, languages and stories about land and nature because close observation of one set of local circumstances leads to divergence from those responding to another set of conditions. The critical elements of knowledge for sustainability can be defined as follows: its local legitimacy; its creation and recreation; its adaptive quality; and its embedded nature in social processes. This knowledge ties people to the land and to one another. Therefore, when landscape is lost, it is not just a habitat or feature. It is the meaning for some people's lives. Such knowledges are often embedded in cultural and religious systems, giving them strong legitimacy. Knowledge and understanding take time to build, though they can rapidly be lost. Writing of American geographies, author Barry Lopez says: *'To come to a specific understanding. . . requires not only time but a kind of local expertise, an intimacy with a place few of us ever develop. There is no way round the former requirement: if you want to know you must take the time. It is not in books.'*[2]

This desire for intimacy with specific landscapes lies deep within us. For some, it involves getting away from the city lights to walk the ploughed fields of winter, crows cawing overhead, or to step across a glacier in the piercing mountain air, or to pause in a sun-pocked clearing deep in mysterious woodlands. For others, it is the intimacy of the daily connection – with cattle that need milking every morning, or the urban park strolled through on the way to work, or the flocks of birds feeding in a back garden. Put together, these link us to a deep and, sadly, often unrecognized connection with whole landscapes. But when these connections are diminished – by modern farming that takes away the hedgerows or trees, or by sprawling suburban settlements – then this intimacy is lost. People stop caring, and the consequences are troublesome. Lopez put it this way: *'If a society forgets or no longer cares where it lives, then anyone with the political power and the will to do so can manipulate the landscape to conform to certain social ideals or nostalgic visions. People may hardly notice that anything has happened.'* When the people who are intimate with the land go, the landscape no longer has any defenders. Again, Lopez identifies the crucial issue: *'Oddly, or perhaps not oddly, while American society continues to value local knowledge, it continues to cut such people off from any political power. This is as true of small farmers and illiterate cowboys as it is for American Indians, Hawaiians and Eskimos.'*

What happens when you ask people in a locality about what is special to them? We use this question as the starting point when interviewing people about their own communities. All too often, outside professionals (whether planners, developers or scientists) begin by asking about problems, and then identify solutions to these problems. As a result, they miss the fine-grained detail about people's connectedness to a place. We find that people focus on two main themes – special things about the community, such as neighbourliness, friends and family, and special aspects of the land, nature and environment. In excluded urban communities, where physical infrastructure is poor, people will often say things like 'we have a strong sense of community', and 'when anyone has a problem, we all pull together to help'. They celebrate tiny spaces of greenery – even though, when placed against a mountain meadow, these spaces are impoverished. They mourn the steady erosion of their community's value through the accrual of graffiti, litter and dumped cars.

In rural communities that are more obviously close to nature, people will select many valued features. In a series of community assessments involving six villages within Constable (the landscape painter) country in the Suffolk and Essex borders, we found that people emphasized more than 130 features special to them in a river valley extending only 20 kilometres by 5 kilometres in area. The most special places are open countryside around settlements, places where people have walked all of their lives and have, in their minds, made their own. Many sites that were named are water features, such as the river, weirs and local streams and water meadows. Special buildings included those with historical interest, together with the schools, churches and village halls that form the social fabric of the region. Put together, these comprise a rich picture of an entire landscape. These are not partial views and knowledge held by a few people, but are widely dispersed throughout the community.

This is not to say that everyone knows their local place intimately. They clearly do not. England is scattered with dormitory villages, populated by commuters working long hours who know their places only at weekends, or when the evenings stretch out in summer. They rarely notice if something is damaged or lost from the local landscape. Even if they do notice, they may not know what to do because they lack social connections. Some, though, arrive from the city with strange values. In the same valley, one wealthy incomer hired two hit men to shoot the rooks nesting in their tree-top rookery on neighbouring land, as they were making too much noise for him. The ensuing scandal within the community did nothing for the birds. They never came back. Nonetheless, it is also true that it sometimes takes 'incomers' with a different perspective on the environment to provoke changes in thinking amongst local people who are wedded, for example, to industrialized agriculture because they know no alternatives. How, then, can we build this necessary literacy about place?

# Building Ecological Literacy

Cognition is the action of knowing and perceiving; therefore, cognitive systems are learning systems. They take in information, process it and change, as a result. A cognitive system coheres – it sticks together different knowledges and still remains a single whole. It goes beyond the *modernist* single code, or even the post-modernist recognition of fractured and multiply different knowledges. It implies synthesis and the capacity to change and adapt. Three decades ago, the Chilean biologists Humberto Maturana and Francisco Varela developed their radical Santiago theory of cognition. They posed the question 'how do organisms perceive?' Their theory centres on the idea that all living organisms continuously bring forth a world – not the one world, but something individually unique arising out of the fundamental differences between the way in which internal neurological processes work and how these processes interact with our environments. We actively construct a world as we perceive it. We are, therefore, 'structurally coupled' with our environment. Such structural coupling describes the way in which a living system interacts with its environment, and these recurrent interactions trigger small changes, adaptations and revisions in the system. Cognition is not a representation, but the continual act of bringing forth a world. The constant dance of cognitive systems, continually shaping, learning and adapting to their environment, thus describes our relationship with nature.

James Scott, in his visionary book *Seeing Like a State*, deploys the Greek term *mētis* to describe *'forms of knowledge embedded in local experience'*. *Mētis* is normally translated as 'cunning' or 'cunning intelligence'; but Scott says this fails to do justice to a range of practical skills and acquired intelligence represented by the term. He contrasts such *mētis* with the *'more general, abstract knowledge displayed by the state and its technical agencies'* by describing 'villagization' in Tanzania and Ethiopia, Soviet collectivization, the emergence of high-modernist cities, and the appalling standardization of agriculture. Failures come when we design out *mētis* because the state rarely makes the kinds of necessary daily adjustments required for the effective working of systems. *Mētis*, Scott says, is *'plastic, local and divergent. . . It is, in fact, the idiosyncrasies of mētis, its contextualities, and its fragmentation that make it so permeable, so open to new ideas'*.[3] What is encouraging is that an increased number of government departments have found the methods and processes to work sensitively with local people; and the recent spread of sustainable agriculture discussed in earlier chapters is partly a result.

Ecological literacy can be created relatively rapidly, and does not necessarily have to have great antiquity. This is what offers us all hope. Farmer field-schools in South-East Asia create new intimate knowledge

about entomology in rice fields; water-user groups develop new under-
standings of the joint management of irrigation water for whole communities; and farmers' experimenting groups in Australia, Europe and North
America develop new ways in which to farm, using few fossil-fuel derived
inputs. This knowledge soon becomes bound up in new rituals and
traditions, which then confer a greater sense of value and permanence. It
would be wrong, therefore, to think of *mētis* as traditional knowledge because
this mistakenly gives the impression that such intimate local knowledge
is unchanging, rigid and unable to adapt. Instead, it is the process of
knowing, and it is central to the idea of ecological literacy.

The idea of the world being full of diverse, parochial conditions, with
each place needing a differentiated approach, does not fit well with the
standardizing approach of industrial development. Modernism is efficient
because it aims for simplification. The central assumption is that technological solutions are universal, and therefore are independent of social
context. Ironically, this is also what makes it appealing – mass production
for us all. In some sectors, it works. Does it matter if the only restaurant
we can visit is the same as those in thousands of other cities around the
world? Yes it does, though we can always choose not to go. But does it
matter if the technology to produce our food is standardized, and
therefore requires coercion in order to encourage adoption by farmers.
Clearly, it does – it matters for farmers because their choices diminish and
their risks increase.

When farmers' conditions happen to be similar to those where technologies are developed and tested, then the technology is likely to spread.
But most farmers experience differing conditions, values and constraints.
When they reject a technology – for example, because it does not fit their
needs or is too risky – modern agriculture can have no other response but
to assume it is the farmers' fault. Rarely do scientists, policy-makers and
*extensionists* question the technologies and the contexts that have generated
them. Instead, they blame the farmers, wondering why they should resist
technologies with such 'obvious' benefits. It is they who are labelled as
'backward' or 'laggards'. The problem, as architect Kisho Kurukawa
indicates, is that *'Technology does not take root when it is cut off from culture and
tradition. The transfer of technology requires sophistication: adaptation to region, to unique
situations and to custom.'*[4]

Modernist thinking inevitably leads to a kind of arrogance about the
social and natural world. It allows us to make grand plans without the
distraction of consulting with other people. It allows us to cut through
the messy and complex realities of local circumstance. Such modernity
seeks to sweep away the confusion of diverse local practices and pluralistic
functions, accumulated over the ages, in order to establish a new order.

This is perceived as an order that brings freedom from the constraints of history, and the promise of liberty. But simplified rules and technologies can never create properly functioning communities. There will always be something missing. Sadly, during the 20th century, we pushed nature and our communities far from equilibrium. Now we need to discover new equilibria by reshaping the world. Barry Lopez says, *'To keep landscape intact and the memory of them, our history in them alive, seems as imperative a task in modern times as finding the extent to which individual expression can be accommodated.'*[5]

The fundamental contradiction of modernity centres on standardization, which goes against the idea of self-made, or *autopoiētic*, systems. For Maturana and Varela, cognition involves perception, emotion and action. We can shape, do and think differently. But modern life has witnessed those with an intimate knowledge of land and landscape being disenfranchised. It has removed their linkages, their structural coupling, their meaning. A world faced by fundamental ecological challenges must therefore be reshaped by collective cognitive action.

A persistent problem is that the dualistic modes of thought go very deep.[6] We have learned them well, and find it difficult to shake them off. Technological determinism is a dominant feature of modernist thought and action, and science and technology are understood as having control over nature, with the solutions to nature's problems lying in cleverer and more sophisticated technologies. At the other end of the spectrum are those who suggest that nature itself is no more than a social construction, with no ecological absolutes or opportunities for technologies to provide any value. In truth, the answer lies somewhere in the middle. We are not separated from nature; we are a fundamental part of a larger whole, and we do have some technological fixes. But we still need clear thinking and theories in order to ensure that we do not imply that by simply joining hands with nature, all will be well.

Nonetheless, from regular use comes accumulated knowledge, and with the knowledge comes understanding and value for local resources. Since these are shaped by the specificities of climate, soil, biodiversity and social circumstances, they differ from place to place. This inherent, ingrained diversity is what we value. It is what gives a place its character and its uniqueness. If we are to protect it, then we have to find new ways of understanding and of creating the collective will to act differently. Niels Röling uses the terms 'beta and gamma science' to describe the need for new forms of interactive design and management in order to help us move away from ecological catastrophe. These terms go beyond alpha science, which tends to be single disciplinary. Röling coins the phrase 'global garden' to reflect the *'conviction that the Earth must be looked upon as a garden tended by human collective action. . . no ecosystem, be it wetland, forest, mountain range, or watershed will continue* to exist or be regenerated unless people deliberately set out to create conditions for it and agree to act collectively to that end'.[7] A key challenge centres on how we can promote such collective action.

## Ideas About the Term 'Social Capital'

The term 'social capital' is used to give importance to social bonds, norms and collective action. Its value was identified by Ferdinand Tönnies and Petr Kropotkin in the late 19th century, shaped by Jane Jacobs and Pierre Bourdieu 70 to 80 years later, and given novel theoretical frameworks by sociologist James Coleman and political scientist Robert Putnam during the 1980s and 1990s. Coleman describes it as *'the structure of relations between actors and among actors'* that encourages productive activities. These aspects of social structure and organization act as resources for individuals to realize their personal interests. As social capital lowers the costs of working together, it facilitates cooperation. People have the confidence to invest in collective activities, knowing that others will do so, too. They are also less likely to engage in unfettered private actions that result in resource degradation. There are four central features of social capital: relations of trust; reciprocity and exchanges; common rules, norms and sanctions; and connectedness, networks and groups.[8]

Trust lubricates cooperation, and therefore reduces the transaction costs between people. Instead of having to invest in monitoring others, individuals are able to trust them to act as expected. This saves money and time. It also creates a social obligation. By trusting someone, this engenders reciprocal trust. There are two types of trust: the trust we have in individuals whom we know; and the trust we have in those whom we do not know, but which arises because of our confidence in a known social structure. Trust takes time to build, but is easily diminished; and when a society is pervaded by distrust, cooperative arrangements are very unlikely to emerge or persist.[9]

Reciprocity and regular exchanges increase trust, and therefore are important for social capital. Two types of reciprocity have been identified. Specific reciprocity refers to simultaneous exchanges of items of roughly equal value, while diffuse reciprocity refers to a continuing relationship of exchange that at any given time may be unrequited, but which over time is repaid and balanced. Again, this contributes to the development of long-term obligations between people, which is an important part of achieving positive sum gains for the environment.[10]

Common rules, norms and sanctions are the mutually agreed or handed-down norms of behaviour that place group interests above those of individuals. They give individuals the confidence to invest in collective or group activities, knowing that others will do so, too. Individuals can take responsibility and ensure that their rights are not infringed upon. Mutually agreed sanctions ensure that those who break the rules know that they will be punished. These rules of the game, also called the internal morality of a social system, the cement of society, and the basic values that shape beliefs, reflect the degree to which individuals agree to mediate their own behaviour. Formal rules are those that are set out by authorities, such as laws and regulations, while informal ones shape our everyday actions. Norms are, by contrast, preferences and indicate how we should act. High social capital implies high internal morality, with individuals balancing individual rights with collective responsibilities.[11]

Connectedness, networks and groups, and the nature of relationships, are the fourth feature of social capital. Connections are manifested in many different ways, such as the trading of goods, the exchange of information, mutual help, the provision of loans, and common celebrations and rituals. They may be one way or two way, and may be long established, therefore not responding to current conditions or subject to regular update. Connectedness is institutionalized in different types of groups at the local level, from guilds and mutual aid societies to sports clubs and credit groups; from forest, fishery or pest management groups to literary societies and mothers' groups. High social capital also implies a likelihood of multiple membership of organizations and good links between groups. In one context, there may be numerous organizations, but each protects its own interests with little cross-contact. Thus, organizational density is high, but inter-group connectedness low. In another context, a better form of social capital implies high organizational density and many cross-organizational links.[12]

Connectivity has many types of horizontal and vertical configuration. It can refer to social relationships at community level, as well as between government ministries. It also refers to connectedness between people and the state.[13] Even though some agencies may recognize the value of social capital, it is rare to find all of these connections being emphasized. For example, a government may stress the importance of integrated approaches between different sectors, but fail to encourage two-way vertical connections with local groups. Another may emphasize the formation of local associations without building their linkages upwards to other external agencies. In general, two-way relationships are better than those that are one way, and linkages that are regularly updated are generally better than historically embedded ones.

# Social and Human Relations as Prerequisites for Improving Nature

New configurations of social and human relationships are prerequisites for long-term improvements in nature. Without changes in thinking, and the appropriate trust in others to act differently, too, there is little hope for long-term sustainability. It is true that natural capital can be improved in the short term with no explicit attention paid to social and human capital. Regulations and economic incentives are commonly used to encourage changes in behaviour, such as the establishment of strictly protected areas, regulations for erosion control, and economic incentives for habitat protection. But though these may change behaviour, they do not guarantee a change in attitudes: farmers commonly revert to old practices when the incentives end or regulations are no longer enforced.[14]

There are quite different outcomes when social relations and human capacity are changed. External agencies or individuals can work with individuals in order to increase their knowledge and skills, their leadership capacity and their motivations to act. They can work with communities to create the conditions for the emergence of new local associations with appropriate rules and norms for resource management. If these succeed in leading to the desired improvements in natural resources, then this has a positive feedback on both social and human assets. When people are organized in groups, and their knowledge is sought, incorporated and built upon during planning and implementation, then they are more likely to sustain activities after project completion.[15] Michael Cernea's study of 25 completed World Bank projects found that long-term sustainability was only guaranteed when local institutions were strong. Contrary to expectations at the time of project completion, projects failed when there had been no focus on institutional development and local participation.[16]

There is a danger, of course, of appearing too optimistic about local groups and their capacity to deliver economic and environmental benefits. We must be aware of the divisions and differences within, and between, communities, and how conflicts can result in environmental damage. Not all forms of social relations are necessarily good for everyone in a community. A society may be well organized, have strong institutions and have embedded reciprocal mechanisms, but may be based on fear and power, such as in feudal, hierarchical, racist and unjust societies. Formal rules and norms can also trap people within harmful social arrangements. Again, a system may appear to have high levels of social assets, with strong families and religious groups, but contain abused individuals or those in conditions of slavery or other forms of exploitation. Some associations can also act as obstacles to emerging sustainability, encouraging conformity, perpetuating adversity and inequity, and allowing some individuals to get others to act in ways that suit only themselves. We must always be aware of the dark side of social relations and connectedness.[17]

We need new thinking and practice in order to develop and spread forms of social organization that are structurally suited to natural resource management. This means more than just reviving old institutions and traditions. More often, it requires new forms of association for common action. It is also important to distinguish between social capital that is embodied in groups, such as sports clubs, denominational churches and parent-school associations, and social capital that is found in resource-oriented groups. It is also important to distinguish between high-density social capital in contexts with a large number of institutions but little cross-membership and high exclusion, with social capital in contexts with fewer institutions but multiple overlapping membership of many individuals.[18]

For farmers to invest in collective action and social relations, they must be convinced that the benefits derived from joint approaches will be greater than those from 'going it alone'. External agencies, by contrast, must be convinced that the required investment of resources to help develop social and human capital, through participatory approaches or adult education, will produce sufficient benefits that exceed the costs. Elisabeth Ostrom puts it this way: *'Participating in solving collective-action problems is a costly and time consuming process. Enhancing the capabilities of local, public entrepreneurs is an investment activity that needs to be carried out over a long-term period.'* For initiatives to persist, the benefits must exceed these costs and those imposed by any free riders in collective systems.[19]

## Participation and Social Learning

The term participation is now part of the normal language of most development and conservation agencies. It has become so fashionable that almost everyone says that it is part of their work. This has created many paradoxes because it is easy to misinterpret the term. In conventional development, participation has commonly centred upon encouraging local people to contribute their labour in return for food, cash or materials. But material incentives distort perceptions, create dependencies and give the misleading impression that local people are supportive of externally driven initiatives. When little effort is made to build local interests and capacity, then people have no stake in maintaining structures or practices once the flow of incentives stops. If people do not cross a cognitive frontier, then there will be no ecological literacy.[20]

The dilemma for authorities is that they both need and fear people's participation. They need people's agreement and support, but they fear that wider and open-ended involvement is less controllable. However, if this fear permits only stage-managed forms of participation, then distrust and greater alienation are the most likely outcomes. Participation can mean finding something out and proceeding as originally planned. Alternatively, it can mean developing processes of collective learning that change the way in which people think and act. The many ways in which organizations interpret and use the term participation range from passive participation, where people are told what is to happen and act out predetermined roles, to self-mobilization, where people take initiatives independently of external institutions.[21]

Agricultural development often starts with the notion that there are technologies that work, and so it is just a matter of inducing or persuading farmers to adopt them. But the problem is that the imposed models look good at first, and then fade away. Alley cropping, an agroforestry system comprising rows of nitrogen-fixing trees or bushes separated by rows of cereals, has long been the focus of research. Many productive and sustainable systems that need few or no external inputs have been developed. They stop erosion, produce food and wood, and can be cropped over long periods. But the problem is that very few farmers have adopted these systems as designed – they appear to have been produced largely for research stations, with their plentiful supplies of labour and resources, and standardized soil conditions.[22]

It is critical that sustainable agriculture and conservation management do not prescribe concretely defined sets of technologies and practices. This only serves to restrict the future options of farmers and rural people. As conditions change and as knowledge changes, so must the capacity of farmers and communities enable them to change and adapt, too. Agricultural sustainability should not imply simple models or packages that are imposed upon individuals. Rather, sustainability should be seen as a process of social learning. This centres upon building the capacity of farmers and their communities to learn about the complex ecological and biophysical complexity in their fields and farms, and then to act on this information. The process of learning, if it is socially embedded and jointly engaged upon, provokes changes in behaviour and can bring forth a new world.[23]

We could think of nature and farm fields as being full of megabytes of information, thereby ensuring a focus on developing the proper operating systems for a new sustainability science. Genetics, pest–predator relationships, moisture and plants, soil health, and the chemical and physical relationships between plants and animals are subject to manipulation, and farmers who understand some of this information, and who are confident about experimentation, have the components of an advanced operating system. This is social learning – a process that fosters innovation and adaptation of technologies that are embedded in individual and social transformation. As a result, most social learning is not to do with hard information technology, such as computers or the Internet. Rather, it is associated, when it works well, with farmer participation, rapid exchange and transfer of information when trust is good, better understanding of agroecological relationships, and farmers experimenting in groups. Large numbers of groups work in the same way as parallel processors, the most advanced forms of computation.

## The Creation of New Commons

We treat nature as property in several different ways. In one setting, nature may be private property, and so only used by a limited number of people. In another, it may be controlled by the state, perhaps on behalf of a larger group of people, or to restrict access by another group. In yet another, nature may be held as a common property. Finally, it may not be controlled or managed at all, and therefore available for use by anyone who wishes. These controls matter because they determine the outcomes for nature.[24] *Common property* or *common pool resources* are technically defined as those used in common by an identifiable group of people, and from which it is too costly to exclude users who obtain individual benefits from their use. A key feature is that they are interdependent systems in which individual actions affect the whole system. If these actions are coordinated, then individuals will enjoy higher benefits (or reduced harm), when compared with acting alone. But if this joint management breaks down, then some may benefit greatly in the short term by extracting all the benefit for themselves. In this case, the likely outcome is damage to the whole system.

There are many types of common resources that are shared by communities of producers and consumers. They include forests and aquifers, fisheries and wildlife, roads and public hospitals, carbon reserves in the soil, and the air we breathe. They exist at different levels of aggregation, from the local to the global. At the local, they comprise irrigation water, forests and grazing lands. At a national level, they include fish stocks in lakes, soil stocks, biodiversity and landscapes. At a regional level, they are manifested in large *watersheds* and basins, such as the North American Great Lakes, the Nile Basin and the North Sea, and in ecosystems that cross national boundaries, such as the Amazon forests. At the global level, they comprise the high seas, Antarctica and the atmosphere. Crossing all of these levels are, of course, food systems. Not so long ago, these systems were solely local; but they have progressively become globalized. These are commons in as much as we all need food, and have a stake and interest in how it is grown or raised.[25]

The origin of modern cooperative action is often dated to 1844 when the Rochdale Pioneers established the first cooperative society in northern England. It led to the establishment of many similar organizations across Europe, providing alternative institutions and services to those available from government. In most developing countries, by contrast, cooperatives have been promoted by governments as instruments of economic development. In India, this phenomenon began with the Indian Cooperative Credit Societies Act in 1904, and most five-year plans since independence have emphasized the roles of cooperatives in agricultural development. By the beginning of the 1990s, there were 340,000 formally registered cooperatives. Many of these, though, seem not to have benefited the poorest.

The problem with conventional cooperatives is well illustrated by Katar Singh's description of the plight of salt miners' cooperatives in Gujarat, which account for 64 per cent of all salt production in India. Most of the value is captured by companies, but licensed cooperatives of salt miners and farmers, locally known as *agrarias*, still survive. In one area, briny water is pumped from more than 100 metres in depth onto surface pans for crystallization, from which the salt is harvested and sold. But the activity is very risky – *agrarias* often fail to strike water, the discharge rate may be variable or suddenly fail, there may be insufficient sunlight, and there are many health risks, owing to the lack of shoes and eye protection. All of these risks are born by the *agrarias*. Thus, these cooperatives, formed by government to improve poor people's social and economic conditions, have failed to do much more than provide organized labour for exploitation. The *agrarias* share of the price that consumers pay for salt stands at a paltry 4 per cent. Salt miners are living, but barely so. Here, connectedness makes little difference in an economic context that is severely stacked against poor people.[26]

But these old-style cooperatives are being replaced by a remarkable new movement of collective-action institutions that are intended to improve people's livelihoods through natural resource management. These are described variously as community management, participatory management, joint management, decentralized management, indigenous management, user-participation, and co-management. These advances in social capital creation have centred upon social learning and group formation in a range of sectors, including watershed or catchment management, irrigation management, *microfinance*, forest management, *integrated pest management* and farmers' research groups. Hugh Ward and I estimate that between 400,000–500,000 new groups have formed in these sectors during the 1990s. Most have evolved to be of small rather than large size, typically with 20 to 30 active members, rising to about 40 for microfinance programmes.[27] This puts the total individual involvement at between 8 and 14 million people – a quite remarkable expansion in social capital and the numbers of ecologically literate people. The real progress towards sustainability has been made by these millions of heroes. They have made collective action and inclusion succeed, and have benefited themselves, as well as the environment.

## Watershed and catchment management groups

Governments and non-governmental organizations have increasingly come to realize that the protection of whole watersheds or catchments cannot be achieved without the willing participation of local people. Indeed, for sustainable solutions to emerge, farmers need to be sufficiently motivated in order to want to use resource-conserving practices on their own farms. This, in turn, needs investment in participatory processes in order to bring people together to deliberate on common problems, and to form new groups or associations capable of developing practices of common benefit. This led to an expansion in programmes focused upon micro-catchments – not whole river basins, but areas usually of no more than several hundred *hectares*, in which people know and trust each other. The resulting uptake has been extraordinary, with participatory watershed programmes reporting substantial yield improvements, together with substantial public benefits, including groundwater recharge, reappearance of springs, increased tree cover, microclimate change, increased common land revegetation, and benefits for local economies. Some 50,000 watershed and catchment groups have been formed in the past decade in Australia, Brazil, Burkina Faso, Guatemala, the Honduras, India, Kenya, Niger, and the US.[28]

## Water users' groups

Although irrigation is a vital resource for agriculture, water is, rather surprisingly, rarely used efficiently. Without regulation or control, water tends to be overused by those who have access to it first, resulting in shortages for tail-enders, conflicts over water allocation, and waterlogging, drainage and salinity problems. But where social capital is well developed, then water users' groups with locally developed rules and sanctions are able to make more of existing resources than are individuals working alone or in competition. The resulting impacts, such as in the Philippines and Sri Lanka, typically involve increased rice yields, increased farmer contributions to the design and maintenance of systems, dramatic changes in the efficiency and equity of water use, decreased breakdown of systems, and reduced complaints to government departments. More than 60,000 water users' groups have been set up in the past decade or so in India, Nepal, Pakistan, the Philippines and Sri Lanka.[29]

## Microfinance institutions

One of the great recent revolutions in developing countries has been the emergence of new credit and savings systems for poor families. These systems lack the kinds of collateral that banks typically demand, appearing to represent too a high a risk. They are therefore trapped into having to rely on money-lenders who charge extortionate rates of interest. A major change in thinking and practice occurred when professionals began to realize that it was possible to provide microfinance to poor groups, and still ensure high repayment rates. When local groups, in particular women, are trusted to manage financial resources, they can be more effective than banks. The Grameen Bank in Bangladesh was the first to help people find a way out of this credit trap by helping women to organize into groups. Elsewhere in Bangladesh, the non-governmental organization Proshika has helped to form 75,000 local groups. Such microfinance institutions are now receiving worldwide prominence: the 50 microfinance initiatives, in Nepal, India, Sri Lanka, Vietnam, China, the Philippines, Fiji, Tonga, the Solomon Islands, Papua New Guinea, Indonesia and Malaysia, have 5 million members in 150,000 groups. Remarkably, these poor people have mobilized US$130 million of their own savings to finance their own revolving credit systems.[30]

## Joint and participatory forest management

In many countries, forests are owned and managed by the state. In some cases, people are actively excluded. In others, some are permitted the right to use certain products. Governments have not been entirely successful in protecting forests, and in recent years have begun to recognize that they cannot hope to protect forests without the voluntary involvement of local communities. The most significant changes occurred in India and Nepal, where experimental local initiatives during the 1980s increased biological regeneration and income flows to the extent that governments issued new policies for joint and participatory forest management in India in 1990, and in Nepal in 1993. These policies encouraged the involvement of non-governmental organizations as facilitators of local group formation – governments realized that they were not good at doing this themselves. There are now nearly 30,000 forest protection committees and forest users' groups in these two countries alone, managing more than 2.5 million hectares of forest, mostly with their own rules and sanctions.[31] Benefits include increased fuelwood and fodder productivity, improved biodiversity in regenerated forests, and income growth amongst the poorest households. Old attitudes are changing as foresters come to appreciate the remarkable regeneration of degraded lands following community protection, and the growing satisfaction of working with, rather than against, local people. There is still a long way to go, though, with an estimated 31 million hectares of forest in India that is still degraded, and state institutions not readily capable of engaging in a participatory fashion with communities.[32]

## Integrated pest management and farmer field-schools

Farmer field-schools have been another significant model for social learning to emerge in the past decade and a half. Integrated pest management comprises the joint use of a range of pest-control strategies (insects, weeds or disease) in a way that reduces pest damage to below economic thresholds and is sustainable and non-polluting. Inevitably, it is a more complex process than just relying on the spraying of pesticides. It requires a high level of analytical skills and understanding of agroecological principles, and it also necessitates cooperation between farmers. Farmer field-schools are called schools without walls, in which a group of up to 25 farmers meets weekly during the rice season to engage in experiential learning. The farmer field-school revolution began in South-East Asia, where research by Peter Kenmore and colleagues on rice systems demonstrated that pesticide use was correlated with pest outbreaks in rice. The loss of natural enemies, and the free services they provided for pest control, comprised costs that exceeded the benefits of pesticide use. The programme of field schools has since spread to many countries in Asia and Africa. At the last estimate, roughly 2 million farmers are thought to have made a transition to more sustainable rice farming as a result. Field schools have given farmers the confidence to work together on more sustainable and low-lost technologies for rice cultivation. It appears, too, that the process of learning is more likely to persist. One study compared farmers in China who had been trained either in field schools or by the calendar spraying methods. Evidence showed that field-school farmers continued to learn in the years after training, whereas conventionally trained farmers experienced no changes in their accumulated knowledge.[33]

## Farmers' groups for co-learning

The normal mode of agricultural research is to experiment under controlled conditions on research stations, with the resulting technologies passed on to farmers. In this process, farmers have little control, and many technologies do not suit them, thus reducing the efficiency of research systems. Farmers' organizations can, however, make a difference. They help research institutions to become more responsive to local needs, and can create extra local value by working on technology generation and adaptation. Self-learning is vital for sustainability and, by experimenting themselves, farmers increase their own awareness of what does and does not work. There have been many innovations in both industrialized and developing countries – though, generally, the number of groups tends to be much smaller than in watershed, irrigation, forestry, micro-finance and pest management programmes.[34]

# The Personal Benefits of Connectedness

Is there any evidence that new forms of connectivity with land that are embedded in local organizations lead to personal change? Ultimately, the fundamentals of the sustainability challenge require us to think differently. I recall being told a story a decade ago by an Indian administrator that captures this idea of the personal frontiers that must be crossed. This administrator had seen the effectiveness of participatory methods else-where, and decided to test them with his own staff. He divided them into two cohorts – those who would receive new training in participatory approaches, and those who would continue to work with local people in the old top-down fashion. He recounted how this experiment had been so effective in the workplace that he had inadvertently found himself treating his driver and his family differently. Once crossed, these boundaries are never revisited.

Gregory Peter and colleagues from Iowa State University, and the sustainable agriculture organization Practical Farmers of Iowa, present compelling evidence of the nature of personal change within households. In most Iowan farms, they say:

> *The division of labour still largely follows gender lines: men do most of the outdoor work, and women support the men's hectic schedules by providing meals at odd hours, doing chores, running the household, going out for tractor parts, and working off-farm jobs – not to mention taking care of the children.*

Using terms developed by Mikhail Bakhtin, they call this monolegic masculinity, which *'mandates a specific definition of work and success'*. But they discovered the emergence of a dialegic masculinity amongst male farmers who were engaged in sustainable agriculture as members of Practical Farmers of Iowa (PFI). They expressed less need for control over nature, were more socially open, were less likely to distinguish between men's and women's roles on farms, and, importantly, were *'more open to talking about making mistakes, to expressing emotions'*.

Monolegic people are individuals who speak and act without acknowledging others, while dialegec social actors take others into account. Industrial farmers were more likely to celebrate long hours and hard work in the form of an ascetic denial of food, relaxation and being with the family. They were also more likely to have a so-called 'big iron' mentality: a love of large machines, which, of course, ooze authority over the land. Sustainable agriculture farmers without these worldviews needed the social connections of being a member of PFI even more, as they often felt isolated and excluded amongst conventional farmers. What this means, in practice, is that farmers who were leaning towards sustainable practices had become another 'sort' of farmer.[35]

Social capital and the experimental capacity of farmers have been developed by the International Centre for Tropical Agriculture in Latin America in groups called Comité de Investigación Agrícultura Tropical (CIAL). Two hundred and fifty groups have been set up in six countries, developing their own individual pathways according to the motivations and needs of farmers. These groups decide upon research topics, conduct experiments and draw upon technical help from field technicians and agricultural scientists. According to Ann Braun, members talk about being *'awakened about their continuous learning process, and losing their fear of speaking out in public'*. There have been many benefits for those involved, comprising more experimentation, easier adoption of new ideas and improved *food security*. Not only do farmers benefit from their experimental findings, they also acquire increased status in the community at large.[36]

Another example of these personal changes comes from central Tamilnadu, where the Society for People's Education and Economic Change (SPEECH) has carefully measured how their partner women's self-help groups developed over a five-year period. Firstly, they found that the incomes and savings of members had increased. More importantly, they found that members' knowledge of banking, income generation, common-property management, health and sanitation, and family planning grew steadily over time. One-year-old groups had a good understanding of income generation and the self-help concept, but less of other issues. Young groups also tended to spend more time in meetings than more mature ones. Members of one-year groups were more tentative in expressing opinions, while those in older groups were more frank.

There were also very important changes within households, with husbands more likely to dominate decisions on household purchases and housing alterations in the early years. After one year, decisions made by the wife occurred only in 6–15 per cent of households. Yet, after five years, decisions were jointly made in 40–60 per cent of households, or by the wife alone in 30–50 per cent of households. SPEECH says: *'Women feel self-confident because their self-help group is backing them and they are recognized in the family and community for their contributions to household income and have more control over family finances than before.'* Older groups are also three times more 'connected up' to other institutions, both within the community and outside. They have better and more regular links with government officials, cooperative societies, police, banks, schools, and other women's groups, including the regional federated body for groups. The younger groups, though, have not yet learned to stand alone.[37]

# The Maturity of Social Capital

This emergence of social capital, manifested in groups and associations worldwide, is very encouraging. It is helping to transform some natural resource sectors, such as forest management in India, with 25,000 forest protection committees, or participatory irrigation in Sri Lanka, with 33,000 groups. Some countries or regions are being transformed. One third of all Australian farmers are members of 4500 Landcare groups, and there are nearly 2 million Asian farmers who are engaged in sustainable rice management.

However, the fact that groups have been established does not guarantee that resources will continue to be managed sustainably or equitably. What happens over time? How do these groups change, and which will survive or become extinct? Some will become highly effective, growing and diversifying their activities, while others will struggle on in name only. Can we say anything about the conditions that are likely to promote resilience and persistence? There is surprisingly little empirical evidence about the differing performances of these groups, though theoretical models have been developed to describe changes in social and organizational structures, commonly characterizing structure and performance according to phases.[38] Some of these focus on the organizational development of business or corporate enterprises, with a particularly strong emphasis on the life cycles of groups. Others focus on the phases of learning, knowing and world views through which individuals progress over time.

Bruce Frank of the University of Queensland and I developed a model to describe how changes in the renewable assets base (natural, social and human capital) affect the performance of managed natural systems, such as farms, forests or fisheries, or regional systems such as watersheds or river basins.[39] Assets can be in one of two states: either in a positive state (and therefore maintained or accumulated), or in a negative state (and therefore degraded). Systems may be producing high levels of desirable outputs, but are doing so by degrading the asset base – for example, because capital is being converted into income, fewer assets remain for future generations. Such systems are productive but, inevitably, unsustainable. Alternatively, systems may have a positive performance or output, but with assets being accumulated. This equates to the more sustainable sector, where systems produce desirable outputs by not degrading renewable assets. We proposed that groups can be found along a continuum characterized by three phases called:

- reactive dependence;
- realization independence; and
- awareness independence.

When groups form, they do so to achieve a desired outcome. This is likely to be in reaction to a threat or crisis, or as a result of the prompting of an external agency. They tend, at this stage, to be looking back, trying to make sense of what has happened. There is some recognition that the group has value; but rules and norms tend to be externally imposed or borrowed. Individuals are still looking for external solutions, and therefore tend to be dependent upon external facilitators. There is an inherent fear of change; members would really like things to return to the way things were before the crisis arose, and before the need to form a group arose. For those groups concerned with the development of more sustainable technologies, the tendency at this stage is to focus on eco-efficiency by reducing costs and damage. In agriculture, for example, this will mean adopting reduced-dose pesticides and targeted inputs, but not yet the use of regenerative components.

The second phase sees growing independence, combined with a realization of newly emerging capabilities. Individuals and groups tend to look inwards more often, and are beginning to make sense of their new reality. Members are increasingly willing to invest their time in the group itself, as trust grows. Groups at this stage begin to develop their own rules and norms, and start to look outwards. They develop horizontal links with other groups and realize that information flowing upwards and outwards to external agencies can be beneficial for the group. With the growing realization that the group has the capacity to develop new solutions to existing problems, individuals tend to be more likely to engage in active experimentation and the sharing of results. Agricultural approaches start incorporating regenerative technologies in order to make the best use of natural capital rather than simple eco-efficiency. Groups are now beginning to diverge and develop individual characteristics. They are more resilient, but still may eventually break down if members feel that they have achieved the original aims, and do not wish to invest further time in pursuing new ones.

The final phase involves a ratchet shift for groups, with greater awareness and interdependence. They are very unlikely to unravel or, if they do, individuals have acquired new world views and ways of thinking that will not revert. Groups are engaged in shaping their own realities by looking forward, and the individual skills of critical reflection (how we came here) combined with abstract conceptualization (how we would like things to be) mean that groups are now expecting change and are more dynamic. Individuals tend to be much more aware of the value of the group itself. They are capable of promoting the spread of new technologies to other groups, and of initiating new groups themselves. They want to stay well linked to external agencies, and are sufficiently strong and resilient to resist external powers and threats. Groups are more likely to come together in apex organizations, platforms or federations in order to achieve higher-level aims. At this stage, agricultural systems are more likely to be redesigned according to ecological principles, no longer adopting new technologies to fit the old ways, but innovating to develop entirely new systems.

The idea of a link between maturity of groups and outcomes raises important questions. Are groups who are endowed with social capital more likely to proceed to maturity, or can they become arrested because social capital is a form of 'embeddedness' that prevents change? Does feedback occur between maturity and social capital? If so, is it positive (for example, success with a new sustainable practice that spills over into success for others, or creates new opportunities for cooperation), or is it negative (such as changes in world views and technology that unsettle traditional practices, erode trust and make existing networks redundant)? Groups and individuals at stage three (awareness independence) appear unlikely to regress to a previous stage, because world views, philosophies and practices have fundamentally changed. But groups at stage one (reactive dependence) are unstable and could easily regress or terminate without external support and facilitation. This raises further challenges for external policy agencies. Can they create the conditions for take-off towards maturity when there is little social capital? How best should they proceed in encouraging transformations that will lead to sustained progress?

## Building Assets for Sustainable Futures

What lessons have we learned from programmes that successfully promote social learning and sustainable natural resource management? The first is that sustainability is an emergent property of systems that are high in social, human and natural capital. When these assets are in decline, then we are retreating from sustainability. Next is the recognition that farmers can improve their agroecological understanding of the complexities of their farms and related ecosystems, and that new information can lead to improved agricultural outcomes. In turn, increased understanding is also an emergent property, derived, in particular, from farmers engaging in their own experimentation, supported by scientists and extensionists, and leading to the development of novel technologies and practices. These practices are more likely to spread from farmer to farmer, and from group to group. These conclusions strongly suggest that social learning processes should become an important focus for all agricultural and natural-resource management programmes, and that professionals should make every effort to appreciate the complementary nature of such social processes with sustainable technology development, and the subtlety and care required in their implementation.

What can be done both to encourage the greater adoption of group-based programmes for environmental improvements, and to identify the necessary support for groups in order to evolve to maturity (and thence to spread and connect with others)? Clearly, international agencies, governments, banks and non-governmental organizations should invest more in social and human capital creation. Building human capital and establishing new forms of organization and social capital are not without their costs. The main danger lies in being satisfied with any degree of partial progress, and therefore not going far enough. As Elisabeth Ostrom has put it, *'Creating dependent citizens rather than entrepreneurial citizens reduces the capacity of citizens to produce capital.'*[40] Of course, group-based approaches are not, alone, sufficient conditions for achieving sustainable natural-resource management. Policy reform is an additional requirement for shaping the wider context, in order to make it more favourable for the emergence and sustenance of local groups. This has clearly worked in countries such as India, Sri Lanka and Australia.

One way to ensure the stability of social connectedness is for groups to work together by federating in order to influence district, regional or even national bodies. This can open up economies of scale, thus bringing about greater economic and ecological benefits. The emergence of such federated groups with strong leadership also makes it easier for government and non-governmental organizations to develop direct links with poor and formerly excluded groups – although, if these groups were dominated by the wealthy, the opposite would be true. This could result in greater empowerment of poor households, as they better draw on public services. Such interconnectedness between groups is more likely to lead to improvements in natural resources than regulatory schemes alone.[41]

But this raises further questions. How can policy-makers protect existing programmes in the face of new threats? What will happen to state–community relations when social capital in the form of local associations and their federated bodies spreads to very large numbers of people? Will the state colonize these groups, or will new broad-based forms of democratic governance emerge? Important questions also relate to the groups themselves. Good programmes may falter if individuals start to 'burn out', feeling that investments in social capital are no longer paying. It is vitally important that policy-makers and practitioners continue to seek ways in which to provide support for the processes that both help groups to form, and help them to mature along the lines that local people desire and need, and from which natural environments will benefit.

There are also persistent concerns that the establishment of new community institutions and users' groups may not always benefit the poor. There are signs that these groups can all too easily become part of a new rhetoric, without fundamentally improving equity and natural resources. If, for example, joint forest management becomes the new order of the day for foresters, then there is a very real danger that some will coerce local people into externally run groups so that targets and quotas are met. This is an inevitable part of any transformation process. The old guard adopts the new language, implies that they were doing it all the time, and nothing really changes. But this is not a reason for abandoning the new. Just because some groups are captured by the wealthy, or are run by government staff with little real local participation, does not mean that all are fatally flawed. What it clearly shows is that the critical frontiers are inside of us. Transformations must occur in the way we all think if there are to be real and large-scale transformations in the land and the lives of people.

## Concluding Comments

In Chapter 1, I wrote of the losses of knowledge about land and nature. If we are to develop sustainable agricultural and food systems – even sustainable economies and societies at large – then we will need to develop new forms of social organization and ecological literacy. Our knowledges of nature and land usually accrue slowly over time. Yet, the immediacy of the challenge means that we must move quickly in order to develop novel and robust systems of social learning. These seek to build up relations of trust, reciprocal mechanisms, common norms and rules, and new forms of connectedness, thus helping in the development and spread of a greater literacy about the land and nature. Great progress on developing new commons is now being made through the actions of hundreds of thousands of groups engaged in collective watershed, water, microfinance, forest and pest management. These collective systems can also promote significant personal changes. Ultimately, the barriers are inside each of us, and large-scale transformations of land and community can only occur if we cross these frontiers, too.

## Chapter 8

# Crossing the Internal Frontiers

## A Fundamental Redesign

Human connectedness to nature has deep roots. For 5 million to 7 million years we walked this earth as hunters and gatherers, entirely dependent upon our knowledge of wild resources, and on our collective capacity to gather plants and catch animals. About 10,000 to 12,000 years ago, we began to domesticate plants and animals. For most of the time since then, the culture of food production was intimately bound up in some form of collective action, and in an intimate knowledge of nature. Where city-states emerged, as in Greece, Rome, Mesopotamia, China, Maya and mediaeval Europe, the number of people no longer needing this intimate connection for their livelihoods grew. But it was not until the advent of the Agricultural and Industrial revolutions, just 200 hundred years ago, that food production in some countries began its drift away from the majority of the population. It is barely two generations since agriculture became industrial, and *modernist* agriculture came to dominate, producing food as commodities. This industrialization of a basic human connection has undermined many things.

So, for 350,000 generations, we care and hunt, use and overuse, harvest and replant, cut and re-seed, and from all this emerges the human condition. The state of the world is an outcome of this relationship. For generations, our effects were globally benign, though not necessarily locally benign. Today, however, we are largely disconnected, and because of that we are less likely to notice when the environment is further degraded, or when valued resources are captured and damaged by others. We are satisfied to know (or, at least we believe we are) that more and more food is being produced. But if we lack the innate connections, we no longer question when environmental and social problems emerge. We do not notice that the extrinsic is damaged at the same time as the intrinsic withers away. Although these breakdowns are symptoms of systemic disarray, there is still hope.

There is a great hero in landscape and community regeneration, and he is the fictional creation of author Jean Giono, resident of Manosque in France for most of his life. In Giono's *The Man Who Planted Trees*, Elzéard Bouffier, shepherd and silent roamer of the hills and valleys of Provence, helps to transform a whole rural system. Giono stands alongside all of the 'greats' of nature and wilderness writing, perhaps surpassing many since his concerns are centred on the connection between land and its people, and on what each can do for the other. According to translator Norma Goodrich, Giono termed his confidence in the future *espérance*, the word describing the condition of living in hopeful tranquillity.[1]

In the fiction, the narrator comes upon Elzéard, who is planting acorns amidst a desert landscape. There are no trees or rivers, houses are in ruin, and a few solitary people eke out a meagre living. '*In 1913, this hamlet of 10 or 12 houses had three inhabitants. . . hating one another. . . all about, the nettles were feeding upon the remains of abandoned houses. Their condition had been beyond hope.*' The unnamed narrator returns 5 years later, then again in 12 years, and finally 32 years after the original visit. During this time, Elzéard continues to plant acorns, and seedlings of beech and birch, and the landscape is steadily transformed. When the forest emerges, then the wildlife returns, the rivers run freely, and the community is regenerated.

*Everything had changed. Even the air. Instead of the harsh dry winds that used to attack me, a gentle breeze was blowing, laden with scents. A sound like water came from the mountains: it was the wind in the forest. . . Ruins had been cleared away, dilapidated walls torn down. . . The new houses, freshly plastered, were surrounded by gardens where vegetables and flowers grew in orderly confusion, cabbages and roses, leeks and snapdragons, celery and anemones. It was now a village where one would like to live.*[2]

This is the glorious key to whole landscape redesign – the creation of places where we would really like to live in *espérance*.[3]

Most of the main principles for redesign are present in this story. There is leadership from a hero, someone willing to take a risk, to do something different for the benefit of others. There is ecological literacy, with knowledge about the particulars of local *agroecology* helping to shape actions. There is the building of social and natural assets as foundations for life and for sustainability. There is also a sense of how long it takes, but just how good are the rewards. However, the shepherd is a loner and achieves change only on a small scale. This new agricultural sustainability revolution will not happen all at once. It will take time, and require the coordinated efforts of millions of communities worldwide. But of one thing we should no longer be in any doubt. This is the way forward, and it offers real hope for our world and its interdependent people and biodiversity.

## An Ethic for Land, Nature and Food Systems

Aldo Leopold's masterpiece, *Sand County Almanac and Sketches Here and There*, was published in 1949, a year after his death. His greatest contribution to us all was the idea of the land ethic. This is a proposal for an ecological, ethical and aesthetic science to shape human interactions with, and as a part of, nature. Leopold's land ethic sets out the idea that the beauty and integrity of nature should be protected and preserved from our actions. Ethics is about limits to freedoms. We are free to destroy nature (and we do), yet we should prescribe and accept certain limits. Leopold sees humans as part of nature, not separated as distant observers or meddlers. In the *Sand County Almanac*, he says:

> We abuse land because we regard it as a commodity belonging to us. When we see land as a community to which we belong, we may begin to use it with love and respect. . . That land is a community is the basic concept of ecology, but that land is to be loved and respected is an extension of ethics.

Such an ethic should be *'a differentiation of social and anti-social conduct'*.

This land ethic implies thinking of land and community as a connected network of parts, which includes us as humans, and in which each element possesses intrinsic rights. There are many different views of this land ethic: some say it is visionary, others that it is dangerous nonsense. But the point remains that most people in industrialized countries still see nature as a bundle of resources that are separate from us. Thus, the land ethic remains radical, more than half a century after it was woven together by Leopold.[4]

In truth, such an ethic is what makes us human – the recognition of, and respect for, these limits. Freedoms are vital, but we have obligations and responsibilities, too. If we accept that we (as global communities) are an intricate part of something, or that something is a part of us (just as our livers or lungs are part of our bodies), it is then absurd to engage in action that endangers a component of the system, since the whole will suffer. The Amazon is not a part of me, so I may destroy it. Yet if I do so, the consequences for the atmosphere are severe, and in the end I will suffer. Leopold understood the connection between economies and nature:

> *I realize that every time I turn on an electric light, or ride on a Pullman, or pocket the unearned investment on a stock or a bond, or a piece of real estate, I am 'selling out' to the enemies of conservation. . . When I pour cream in my coffee, I am helping to drain a marsh to graze, and to exterminate the birds of Brazil. When I go birding or hunting in my Ford, I am devastating an oil field, and re-electing an imperialist to get me rubber.*[5]

These choices matter in today's food system. Each time we buy food, our choices make a difference to nature and communities somewhere – though there is perhaps a danger of overstating the power of consumers in the face of structural economic constraints. We are connected within a much larger system, and we can make these connections work to the good – if we wish. Albert Howard was one of the most influential of British scientists to take a holistic view of the connections between nature and people. He spent 26 years in India, and developed the Indore Process in which modern scientific knowledge was applied to ancient methods. He called for a restoration of agriculture based upon an improvement to the health of the whole system, saying that:

> *The birthright of all living things is health. This law is true for soil, plant, animal and humans: the health of these four is one connected chain. Any weakness or defect in the health of any earlier link in the chain is carried on to the next and succeeding links, until it reaches the last, namely us.*[6]

What do we need to do differently? Perhaps the most compelling of Aldo Leopold's essays is a short, but brilliant, piece entitled 'Thinking Like a Mountain', in which he details the relationship between the wolf, deer and mountain in Arizona. He first recalls his own shooting of a mother wolf caring for a tumbling pack of cubs: *'in those days, we never heard of passing up a chance to kill a wolf'*, and then mourns their loss and his earlier lack of understanding. He goes on to describe the consequences of eliminating the wolves; without them, the deer expand too greatly in numbers, and the mountain loses all its vegetation. In the end the whole system collapses.[7] He says: *'Only the mountain has lived long enough to listen objectively to the howl of the wolf. Those unable to decipher the hidden meaning know nevertheless that it is there, for it is felt in all wolf country, and distinguishes that country from all other land.'* These interconnections are true, though, of all lands, and are again something that Leopold foresaw, echoing Thoreau's phrase of almost a century earlier: *'In wildness is the salvation of the world. Perhaps this is the hidden meaning in the howl of the wolf, long known among mountains, but seldom perceived among men.'*[8]

Leopold would feel at home in today's ancient beech and fir forests of the Carpathian Mountains. These are home to the largest numbers of wolves in Europe, and are testament to the principles of ecological balance and diversity. The Carpathian range stretches 1500 kilometres in a giant elbow from Austria in the west, via the Czech Republic, Poland, Slovakia, Hungary and the Ukraine, and finally to Romania. About half of the area is forest, and the rest flower-rich meadows and valley-floor farms. More than half of the Carpathians are in Transilvania, the region of Romania best known for the fictions of Dracula and werewolves. Today, these Romanian forests contain red and roe deer, wild boar and chamois, and Europe's largest concentration of carnivores – 5500 brown bears, 3000 wolves and 3000 lynx.

Walking through these grand forests, you would not know it, for the predators are mostly mysterious. Some bears have become notorious locally in the city of Brasov. In the Racadau neighbourhood, where harsh tower blocks march in ranks to the forest's edge, habituated bears come down to ransack the garbage on summer nights. Local people seem habituated, too, watching calmly from just a few metres away. Some worry that, one day, there will be a serious incident, and sentiment will turn against the bears. When we walk the forest edge, as dusk falls and the heady scent of resin is in the air, the bears seem no more than distant myths. In local mythology, the forests are also special. They are a friendly hiding place, a protection from enemies, and a part of everyone. Romanians have a saying: *'The forest is our brother.'* The Carpathians are still farmed as they have been for centuries, with small valley farms, and livestock are herded on the common mountain meadows for the whole of the summer. Each year, shepherds lose a few sheep to bears and wolves, some 10 to 20 per flock. To date, though, there has been reasonable balance, and shepherds earn a living despite the dangers. The wolves keep down the numbers of deer, without which the trees would suffer. Tree damage in Bavaria, where there are no wolves, is ten times greater than in Transilvania.

There is something very significant about the Carpathians that goes beyond the quirky behaviour of town bears or the distant howling of wolves. Cultural traditions still persist in a modern world, the landscape is patchy and diverse, and nature coexists with people. But most people are still very poor, and so the core question is whether there can be sustainable economic and social development without throwing away all that is culturally valuable and distinctive. Will Romanians tread the same path that is followed over much of the industrialized world?[9] I am not suggesting that all landscapes should look like the Carpathians, or that substantial tracts of agricultural land should be converted to forests or set aside for nature conservation. I do believe, though, that it is possible to have food-producing systems that complement and enhance nature. Nature, after all, still exists on farms and in fields. Today, there is growing confidence that we can, indeed, make the transition directly to sustainable and productive agricultural and food systems that protect and use nature. This is such a significant break from the recent past that the movement may become another agricultural revolution.

## Thinking Like a Wolf

Our old thinking has lamentably failed the rest of nature, and it is in danger of failing us, too. Today, one in four mammals and one in eight bird species face a high risk of extinction. Some 9 per cent of the world's trees, 8700 hundred species, and 12 per cent of all plants, 34,000 species, are threatened. Species are disappearing at a rate 100–1000 times faster than before humans diverged from apes.[10] Each year, the world is losing at least 1000 species. The wolf has gone, ecosystems are disturbed, and now the mountain is slipping away, with soils clogging rivers and damaging distant ecologies. We are in desperate need of a new way of thinking that reintegrates people and nature. Conservation biologist David Orr of Ohio's Oberlin College put it this way:

> Now we have to learn entirely new things, not because we have failed in the narrow sense of the word, but because we succeeded too well. . . What must we learn? We must learn that we are inescapably part of what Leopold called 'the soil-plant-animal-man food chain'. . .which is to say we must embrace a higher and more inclusive level of ethics.[11]

Thinking like a wolf and a mountain implies a great change to both practice and ethics.

We have to ask – could we make a difference if we change the way we think and act? It is not enough to count and catalogue the richness of the world's biodiversity, while at the same time watching its inexorable depletion. Thinking like a mountain means adopting a new ethic to define our human interactions and connectivity with nature. David Ehrenfeld of Rutgers University and first editor of the journal *Conservation Biology* asks: *'What are we really accomplishing? What are we really changing?'* Of species and ecosystems, he also queries: *'Are we improving health? Are we saving lives?'*

The answer is, of course, yes we are doing a lot, but by no means enough. More of the same thinking will not help us get out of the current crisis. Ehrenfeld says:

> *In the process of deluding ourselves into thinking that our science alone can fix the world, we are likely to end up, at best, utterly ineffectual or at worst doing a good deal of damage. . . Success will most likely come to those who knock down the walls around their expertise (painful as this may be), share their knowledge with the community, and learn from it in return.*[12]

In other words, we will not succeed unless we find better connectivity between scientists of different disciplines, between scientists and local people, between communities and politicians, and between story-tellers and reductionists. Such connectivity implies learning, but does not guarantee it. We are going to need to find new ways of learning about what works and what does not. Time, though, is running out.

There is a simple truth in the stories throughout this book. From the forests of Guatemala to the coasts of England, from the drylands of India to the vegetable valleys of Australia, it has taken individuals with courage and motivation to think something new, something so different that they break all the apparent rules. They step outside a paradigm; they cross a frontier. In doing so, they create new possibilities for all of us. Peter Senge, author of *The Fifth Discipline*, rightly identifies the key question:

> *When things are going poorly, we blame the situation on incompetent leaders, thereby avoiding any personal responsibility. When things become desperate, we can easily find ourselves waiting for a great leader to rescue us. Through all of this, we totally miss the bigger question 'what are we, collectively, able to create?'*

He also notes: *'Nothing will change in the future without fundamentally new ways of thinking. . . To think that the world can ever change without changes in our mental models is folly.'*

We are often resigned to think that we cannot have any influence in the world; yet, we can if we only think differently. We can choose food with the environment and animal welfare in mind. We can buy locally to save on carbon emissions. We can visit our local special places and care for them. Above all, we should believe that what we do matters. In the process, we may discover new meanings in the world and in our communities. We may just be able to find a way to save this one planet of ours. David Suzuki says: *'We are still settlers on this earth.'* Settlers have enormous responsibilities on the frontier. They have to bring the best of the past, yet not allow old thinking to get in the way of new requirements.

Crossing these frontiers is tough. There is no simple cookbook or set of instructions, as the process must be different for each of us and is, therefore, adapted to local circumstance. We can, however, take heart from those who have already begun to write a new story – in the coasts of England, forests and fisheries of Japan, cotton communities of Australia, drylands of India, mountains of Pakistan, hills of Kenya and gardens of New York City.

## The Environmental Designer

The low-lying fields and estuaries of eastern England are under a double threat from the sea. The land has been sinking since the glaciers retreated from northern Britain at the end of the last Ice Age, and thermal expansion of the oceans resulting from climate change is pushing up sea levels. For about 1000 years, local communities have actively protected their coasts by building sea walls strong enough to repel the highest of tides and the severest of storm surges. They have also relied on salt marshes to absorb the energy of the sea. These salt marshes pay – a sea wall with no salting in front of it costs UK£5 million per kilometre to construct, but only one tenth of that if there is a salting. For a variety of reasons, though, salt marshes are disappearing. They are squeezed against sea walls, damaged by pollutants, and drained for farmland and housing. So, we are faced with a difficult choice. Do we continue to invest in repelling the sea, with costs likely to spiral, or do we step outside 50 generations of thinking and look at the landscape in a radically different way?

John Hall of the Essex Wildlife Trust recently did exactly this, and he has big plans for one branch of the Blackwater Estuary. This is a land and seascape of massive skies that seem to stretch forever, whether you stand on the sea wall by basking snakes on a summer's day, or cower behind it in the teeth of a cold, steel-grey winter's gale. This coastline is home to hundreds of thousands of wetland birds, including Brent geese, dunlin, knot, shelduck and redshank. The fisheries amongst the marshes contain important stocks of oysters, cockles, herring, bass, mullet and eels. Fronting the north bank of the Blackwater Estuary in Essex is Abbotts Hall farm, until recently a 280-*hectare* conventional arable farm. The farmland is protected by a 2-metre seawall, on the river side of which are remnants of salt marsh, a once common habitat. The farm itself dates back at least to the *Domesday Book* survey of 1085. In 2000, with the support of several organizations, the Essex Wildlife Trust purchased the farm with a grand design in mind.[13] This is highly productive farmland, yet John and colleagues plan to punch five breaches into the sea wall and allow salt-water irrigation to create 120 hectares of new salt marshes, coastal grazing, reed beds and saline lagoons. The remainder of the farm will be devoted to sustainable agriculture methods, including the reinstatement of hedge-rows, ditches, copses and field margins.

The idea is alarming to some people, who see the sea as the enemy of the land.[14] Yet, what will happen when this change in landscape occurs? Previously, the farm did one thing, and one thing well. It produced arable crops. But now it will do many things, with rare habitats and sustainable farming methods on land, and new habitats for oysters and fish in the sea. The extra salt marsh could benefit the village of Salcott further up the channel, since high tide would be substantially lowered, thus reducing flood risk. As the sea level rises, is it not better that it is used positively to create new habitats that complement other land uses? This diverse landscape will help to do exactly that. The Abbotts Hall project will also be a practical example of coastal realignment that scientists and local people can touch and feel, thus bringing reassurance about sustainable coastal defences.

## The Fisherman and The Poet

There is no greater contrast than between the coastal flats of eastern England and the mountainous landscapes of Japan. Our heroes come from Miyagi and Iwane Prefectures, and they connect forests and the marine environment in the Okawa *watershed*. At the top of the mountains there were deciduous forests, once valued for charcoal and fuel wood; in the estuary in Kesennuma Bay there were famous beds of oysters and edible seaweed known as *konbu* and *nori*. But over the years, the traditional forests were cut down or replaced with conifers, wetlands were reclaimed near the river mouth, the river itself was channelled with concrete, and modern agricultural methods were adopted.

The first sign that something was wrong occurred when the oyster fishery collapsed three decades before the end of the century, closely followed by large changes in seaweed populations. It seems obvious that changes in water quality, combined with variable flows, might have had an effect at the estuary end of the system. But at the time, it took the perceptive observation of a fisherman named Hatakeyama to do something about it. At the end of the 1980s, he saw that there was a connection between the sea, forest and river, and he began to talk to people. He travelled up the watershed – a fisherman out of water. He encouraged the other fishermen of Karakuwa town to organize and reflect, and then begin discussions in upstream villages. Residents of upland Murone agreed to cooperate, and this led to the fishermen themselves providing money and labour for reafforestation of the watershed with deciduous trees. Hatakeyama met people in order to talk about common problems, exchange information and share in replanting. They say the fisheries are now more productive; perhaps more importantly, the fishermen have, with remarkable foresight, become educators. They invite schoolchildren from the top of the mountains to visit the bottom, and take their children upstream to plant trees.

The forests have now been gloriously renamed Forests of Oysters. They are so successful that similar activities are being attempted elsewhere in Japan. The Okawa poet Ryuko Kumatani captures the deep connections with the following poem:

> *The Forest lives a long way off the Sea.*
> *The forest, regarded as blessing of the heaven,*
> *Is nurturing a love from far away, longing for the Sea.*
> *Forest is the lover of the Sea.*[15]

What these fishermen, foresters, villagers and poets have done is to claim a whole watershed as a common. It is a region in which they each act to make a difference. Effectively, this has revived an important Japanese tradition – that of the common lands, or *iriaichi*. Each of Japan's 70,000 traditional villages once had carefully controlled and managed commons, with horizontal *kumi* associations to manage them and set locally appropriate rules and norms. Margaret McKean indicates: *'These thousands of villages managed their common lands for several centuries without experiencing a single tragedy of the commons. . . I have not yet turned up an example of a commons that suffered ecological destruction while it was a common'.*[16] It is only in recent times that there has been steady attrition, with commons appropriated by the state or sold off for alternative uses. For centuries *'villagers themselves invented the regulations, enforced them, and meted out punishments, indicating that it is not necessary for regulation of the commons to be imposed coercively or from the outside'.* Now, these traditions, rights and responsibilities are being reclaimed in the Forests of Oysters.

## The Cotton Women

Dalby is a small linear town on a distant crossroads on the Darling Downs in eastern Australia. Like many similar rural settlements, the whole community and economy is under threat. Farming seems to get tougher year on year, businesses struggle, children must travel further to school, and existing associations seem somehow tired and inappropriate for the challenges that modern society brings. Cotton is big on the Downs, but it, too, is struggling against growing pest resistance and increasing environmental degradation. However, farmers are beginning to get organized to share ideas and practices for novel pest management, and there are 350 families in the growers' association. There is some progress towards sustainability. On the Jimbour Plain, Carl and Tina Graham reflect on the changes, by saying: *Ten years ago, if you saw your neighbour spraying, we'd go out and do it, too.'* Today, they are scouting the fields, using trap crops, managing beneficial insects and using natural viral pesticides. They are creating a mosaic landscape so that sorghum can build up parasites, or lucerne can benefit the cotton. They know there is much to do, and say: *'We still have a lot of learning to do.'*

But something else is changing, too, and it involves relations between men and women in the farm community. On a blisteringly hot noontime, I meet with the Women in Cotton group in Dalby. It is led by Catrina Walton and has 60 members. They convened for the first time in 1997 to talk about the pesticides used in cotton cultivation. One says: *'We found it so powerful, just to get us all together.'* They meet once or twice a month, sometimes for discussion, or to hear talks from external professionals of their choice. They organize farm visits for several hundred children each year. The benefits of the group seem to centre on two things. The first is the value of the meetings. Says one member: *'You feel safe; you don't have to tread carefully with your words.'* Another member explains: *'Social networking is one of the greatest things I get out of this group.'*

The second, though, is more subtle, and this is about changed relations within families. Women say that they did not know enough in the past to ask sensible questions, and mothers tended to be pushed into the background. But now there is greater understanding in families, improved communications and more joint decision-making. One member says: *'It makes for a better marriage.'* Back at Jimbour, Carl says: *'When I come home from the paddock I get asked heaps of questions, and we interact more.'* Tina says: *'Women feel more involved. Now I have ideas for improvement, and can answer questions.'* The women themselves are adding productive value to the system. They read reports, help with marketing and learn about pests and predators. Men tend to lack the social networks that women develop, and these networks help to spread good practice and ideas. But it is not easy. One male agronomist arrived at a recent meeting to give a talk and, in front of 50 women, said: *'Oh, so there's no one here yet.'* Together, though, women and men are slowly redesigning their farm systems, making them more sustainable and productive, and they are doing so by crossing a multitude of personal, family and community frontiers.

## The Land Without a Farmer Becomes Barren

Far from the cotton plains, but in a similarly dry and challenging environment, women and men in southern India are redesigning their landscapes and communities. Farmers in central Tamilnadu live in a rain shadow of the western Ghats, an area known for acute droughts and erratic monsoons. They have a saying: 'The land without a farmer becomes barren – thaan vuzhu nilam thariso.'[17] When I first walked to the village of Paraikulum, an unassuming settlement of 60 households several kilometres from the nearest road, the 50 hectares of land above the village were barren. Villagers could remember a time when they were cultivated, but over the years conflicts and land-grabs had undermined cooperation. This area is so dry that it is home to India's match and firework industry. Yet, when it does rain, the water simply rushes off of the barren land to join distant rivers, and a valuable resource is wasted.

Paraikulum, though, is lucky. It is one of 47 villages in the district of Virudhunagar with whom a group called the Society for People's Education and Economic Change collaborates. John Devavaram, Erskine Arunothayam and Rajendra Prasad and their colleagues began their work here in the mid 1980s. Their approach has been to help form self-help groups, or *sanghas*, and to build the social and human capital of the area so that environmental improvements can be made and then sustained. The effect on the landscape and community has been remarkable. After the Paraikulum women's *sangha* was formed, Pandiyammal, Laxmi and neighbours decided to recover the barren land through careful planning of physical and biological water-harvesting measures. They redesigned the fields, so that water would be used effectively.

Now when it rains, the water is channelled, collected and ponded, and seeps into the ground to replenish aquifers. This has produced a double benefit. Firstly, the watershed turned green within three years, as crops could now be grown there. Secondly, enough water was collected in the tank for the community to irrigate the tiny 12-hectare patch of wetland rice close by the village for an extra season each year. With existing resources (after all, the water had been there), they now produce an extra 30–50 tonnes of rice each year. Increasing cropping intensity on small patches in this way offers great hope for farmers of the drylands in many parts of the world. But it requires community organization and motivation. The upper watershed is now unrecognizable. When I first walked across the dusty scrubland, it was good for only goats and the collection of scraps of firewood. The soils were sandy and barely capable of sustaining grasses and shrubs, let alone crops. Today, there are fruit trees on the field boundaries, and fields full of swaying finger millet or creeping groundnut. All of this comes about through collective action, led, in this case, by the local women's group, who had, in turn, been helped by an active and enlightened local group. Put the farmers on the land, and it becomes productive again.

## People in the Mountain Desert

North of these drylands lies an even more challenging environment, the mountain deserts of what is now northern Pakistan. The Karakorams, Pamirs and Hindu Kush, at the western end of the Himalayan range, contain two of the highest mountains of the world – the prosaically named K2 and spectacular Rakaposhi. It is too far inland to catch the monsoons that yearly sweep up the Indian Ocean. But here people have carved out fields from rocky fans, channelled water from distant glaciers, and turned barren and hostile mountainsides green over the centuries. High above, alpine pastures support endangered ibex and snow leopards. The feats of imagination, combined with locally adapted engineering, are extraordinary. Without these people of the Hunza Valley, of Baltistan and Chitral, there would only be desert.

The appealing idea of wilderness, where you can stroll with backpack and remain in touch through satellite communications, is almost undermined by these harsh environments. In winter, temperatures plummet tens of degrees below zero, and people may not venture outside for a couple of months. The growing season is short and sharp, and in high summer it can reach 40 degrees Centigrade or more. The infrastructure is poor, and for centuries people have lived at subsistence level. There was poverty, hardship and inequity. Over the past 20 years, though, the physical and natural environment of these communities has been gradually transformed, mainly due to the work of the Aga Khan Rural Support Programme. Members of the programme realized that huge potential rested in these communities, if only village organizations and, in many cases, women's organizations could be formed to address collective problems and opportunities. They have helped to form 1800 village organizations with 75,000 members, and another 770 women's organizations with 26,000 members. These organizations have built new link roads, reclaimed lands for farming, constructed irrigation channels from glacial melts, and introduced new farm and post-harvest technologies. The World Bank says:

*The programme's most significant contribution is not so much in the number of trees planted or additional area irrigated, but in attitudinal change; people begin to believe they can influence and achieve their development agenda; and the effect of giving women knowledge and self-esteem may outweigh all [other benefits].*[18]

It is easy to underestimate the significance of these achievements. I once flew in one of the Aga Khan's helicopters from Chitral in the west, across the Shandur Pass, location of a famous annual polo match at some 4200 metres, and down the 200-kilometre long Ghizer valley towards Gilgit. The time of year was July, but the hillsides were brown and grey rock, picked out with tiny patches of green around villages, or yellow waving barley and wheat. You cannot take in the scale. The helicopter flew at about 4500 metres for safety, but the valley stretched up another kilometre. And down, too – poring over a map in Gilgit, I found out later that the top of this valley was above 6000 metres, the floor at some 1500 metres. The Ghizer valley is 4.8 kilometres deep, and is just one of many in the region. It would swallow the Grand Canyon in the US. But before we get too romantic about these landscapes of villages and fields clinging to hillsides, it is important to be reminded of the deep poverty, lack of economic and educational opportunity, and unremittingly tough climate. Nevertheless, if the people of the villages of Hunza can build their own schools for girls, the first ever known, or can reclaim a glacial fan and share the new farmland equally, then elsewhere in the world, should we not be wondering if we can do the same? Perhaps we have something to learn from them.

## The Ministry Officials in Kenya

For too long, Africa has been dismissed as being somehow unable to engage in patterns of development that aid people and environments. Food production lags well behind the rest of the world – it is 10 per cent less per person than 40 years ago. Yet, such a view misses a critical truth. Despite the difficulties, there are many who have crossed the frontier, and who, too, have important lessons for the rest of the world, if only we would listen. Some of these new thinkers are in governments, and their successes are even more remarkable. One example comes from Kenya, where a decade ago Ministry of Agriculture officials came up with a new way of protecting soils and communities. Kenya has a long history of state intervention in both soil and water conservation and land management. For five to six decades, farmers had been coerced or paid to adopt soil conservation methods. These methods require coordinated action at a catchment level, and state suspicion of local people's lack of knowledge meant that these decisions imposed upon farmers. In the long run, this approach simply does not work, as people do not maintain the structures over which they feel no ownership. By the end of the 1980s, it had become painfully clear that the conventional approach to soil and water conservation was not conserving soils.

However, a group of soil conservation officials, led by J K Kiara, Maurice Mbegera and M Mbote, recognized that the only way to achieve widespread conservation coverage was to mobilize people to embrace soil and water conserving practices on their own terms. All financial subsidies were stopped, and resources were allocated, instead, to participatory processes, good advice and training, and farmer trips. The catchment approach was adopted in 1989, and was seen as a way of concentrating resources and efforts within a specified catchment, typically 200–500 hectares, for generally one year, during which all farms are laid out and conserved with full community participation. Small adjustments and maintenance are then carried out by the community members themselves, with the support of extension agents.

But these participatory methods imply shifts of initiative, responsibility and action to rural people themselves, and this is not easy for government officials who are used to getting their own way. Moreover, cross-disciplinary teams are drawn from various government departments, such as those with responsibility for education, environment, fisheries, forestry, public works, water and health, in order to work together with local people. A catchment conservation committee of farmers is elected as the institution responsible for coordinating local activities. Quietly, and with little fuss, some 4500 committees have been formed over the decade, and by the late 1990s, about 100,000 farms were being conserved a year. This was more than double the rate, and with fewer resources, than in the 1980s, a time when much of the long-term benefits of conservation were in doubt because of local disapproval of the imposed approach.

The process of implementing the catchment approach itself has, of course, varied according to the human resources available and differing interpretations of the degree of participation that is necessary to mobilize the catchment community. Some still feel that farmers should simply be told what to do. This approach inevitably fails. Others do not invest enough time in developing relations of trust. But where there is genuine participation in planning and implementation, the impacts on food production, landscape diversity, groundwater levels and community well-being are substantial.

# The Magic Gardeners

This book has mostly been about the redesign of rural landscapes and communities. Yet, there will soon be more urban than rural people. Urban people already play two potentially critical roles in rural redesign: by buying food and by visiting rural landscapes and wild areas. There is, though, a third opportunity that is helping to encourage the internal transformations necessary for a more sustainable world. This is urban gardening. At first glance, this might appear a marginal activity, even the word 'gardening' implies a leisure activity for those with time to spare. But urban gardeners, both individual and collective, are also part of this new agricultural revolution. In developing countries, it is already common for large numbers of urban families to be directly engaged in food production. It has been estimated that 100–200 million urban households farm in the city, providing food for some 700 million people.[19] In some Latin American and African cities, up to one third of vegetable demand is met by urban production; in Hong Kong and Karachi it is about half; and in Shanghai more than four-fifths. In Cuba, it is a central part of the whole country's *food security*. In industrialized countries, far fewer people grow their own food. For those who do, though, it is an increasingly important source of psychological well-being.

In the UK, home gardens and allotments used to be vital sources of food. During the early 20th century, there were 1.5 million hectares of allotments producing about half of all fruit and vegetables consumed domestically. Today, the area has fallen to less than 15,000 hectares, eroded by dying interest and growing urban development. Nonetheless, the 300,000 families who garden these allotments are estimated to produce in excess of 200,000 tonnes of fresh produce each year, worth UK£560 million pounds.[20] In the US, the National Gardeners Association estimates that 35 million people grow food in their back gardens and allotments, annually producing 6 billion kilogrammes of food worth US$12 billion to US$14 billion per year. Elsewhere in Europe, urban gardening is hugely popular in Germany, with 50,000 Berliners growing their own food, and in Russia, where urban and rooftop gardens are now common.

But why do people bother, when modern agriculture is so successful at producing large amounts of food for industrialized economies? For some, the food is vital, particularly in economies in transition where food supply is still insecure. But for most people, it is primarily a matter of psychological well-being and improved social relations, supplemented by the bonus of nutritious and healthy food.[21] Urban gardens are special places in the city, oases of tranquillity and repose. They tend, though, to be invisible amongst the pace and dynamics of the city, and so are easily ignored and undervalued. One of the world's most intense urban spaces is surely New York City. Most people feel that they know New York, even if they have not been there. Yet, in amongst the skyscrapers another transformation is underway. This is the GreenThumb movement of community gardens, led heroically for a decade and a half by Jane Weissman. GreenThumb is New York City's community gardening programme, promoted from within the municipal authority, and aimed at turning vacant lots blighted with rubbish, rats and abandoned cars into thriving community gardens.

Derelict lands have been transformed, according to Jane Weissman, into 'safe, thriving and productive oases of green', and now about 20,000 households are actively involved in managing 700 community gardens.[22] These gardens produce US$1 million worth of fruit and vegetables each year, and also revitalize neighbourhoods, foster community pride, provide safe meeting places for local youngsters and seniors, and offer job training. But perhaps more importantly, they are people's little patches of wildness in the city. Bertha Jackson, of the 127th Street Block Association in Central Harlem, says: 'This is the beauty. Yearly we got two or three bushels of peaches from the tree. People have come from near and far for Harlem grown peaches from our garden tree. The peach that grew in Harlem.' Nearby Mary Sciales says:

> Our community garden was created by students, staff, neighbours, community workers and environment groups. Together we. . . have improved the environment, which has made East New York a more beautiful place to work and live. Flowers bloom, vegetables are harvested, the smell of barbecues fills the air and the students learn. They enjoy learning outside. . . and our gardens are an oasis of beauty in the deserts of urban decay.

Donna Armstrong's recent study of 63 community gardens in New York State shows just how valuable these are to local people. She found that they changed local residents' attitudes to their own neighbourhood, resulting in improved care for properties, reduced littering and increased pride in the locality. She also found that community gardens promoted social cohesion and encouraged people to work cooperatively on a range of local needs, such as shared child-care. In short, they improve social capital and personal well-being. Those engaged in community gardening have crossed another frontier, with four out of five saying that their mental health has improved.

But not all is well. Though Jane Weissman was honoured in 1998 with an appointment to the People's Hall of Fame, the city authorities do not like the gardens. They do not recognize their value to local people, and have transferred all responsibilities away from GreenThumb to the housing department. They want the sites for buildings, and intend to have them. They are missing the huge significance of this revolution. Lydia Brown of East Harlem says:

> *If you had seen that trash-filled lot, you'd have said it would take a miracle to make a garden. In time and with much hard work we accomplished the impossible. Now we have something beautiful to look at — flowers, fruits and vegetables for the community. When people walk by, they compliment the garden. One surprised person said 'it's magic'. So we called the garden the Magic Garden. But, in reality, the magic is within us.*[23]

## Connecting up the Promising Cases

These cases, and the stories of transformation that form the backbone of this book, are still in the minority. Yet, time is short, and the challenge is simply enormous. The time has come to take seriously the opportunities offered by this revolution in agricultural and food systems. There is already promising evidence that it can work, and we should wonder: what would be achieved if we all realized that another future is possible? The state of the world and its communities is at stake. Sustainable agricultural and food systems can right many wrongs; but, of course, salvation will not come from these sources alone. Ultimately, if there is to be systemic change centred on both individual transformations in thought and collective changes in action, then it is a question of politics and power.

As I have indicated, everyone is in favour of the idea of sustainability, yet few seriously go beyond fine words. Only two countries in the world, accompanied by some progressive provinces, states or counties within countries, have explicit national policies for sustainable agriculture to encourage transitions in the whole landscape. There are also promising advances within some sectors, and arising out of specific programmes in others. Yet, such sustainable systems of food production are clearly good for whole nations. They help to produce food in an efficient way. They are fairer and more equitable, offering real opportunities for the hitherto poorer and marginalized groups to lead at least reasonable lives. They help to protect nature and the vitally important, but often hidden, services upon which we so depend.

When governments or policy-makers hesitate, then it will be necessary for those who believe in this vision to organize and show the strength that comes from collective will. Most people have something to gain by supporting a sustainable agricultural revolution. But some will feel they have too much to lose: economic power because a product is no longer required; political power because existing support erodes as new forms of social organization emerge; or personal power because the sun is setting on an old idea. Yet, as we have seen, those who progress beyond the internal frontiers originate from many places and perspectives. There is nothing to support a contention that only certain sorts of people are able to make such a transformation. Equally, there is nothing to say that all individuals will. This millennium offers all kinds of opportunities to make the future sustainable for life on Earth. Putting the culture into agri-culture and the wonder and magic back into nature, together with connecting up food systems along agroecological, knowledge-based and community-oriented principles, can, I believe, help in these wider transformations.

This means that we will need to develop entirely different ways of thinking about the value of food, and the additional services or side effects that come from sustainable agricultural systems. Public money that is used to subsidize farmers in industrialized countries, or to support agricultural development in developing countries, should be targeted towards both the provision of *public goods* and services, and the creation of fairer, more equitable systems of production. Some private markets, such as for *carbon sequestration*, will provide further income to farmers engaged in sustainable agriculture. Further sources of income will come from ecotourism. Farming must reorient itself as a *multifunctional* activity, with diverse cultural and environmental connections. An important step is the radical reform of national agricultural policies. Without such change, advances seen to date will stay small scale and parochial. With it, a new direction towards a multipurpose and sustainable agriculture, tied closely to cultures and communities, could become increasingly mainstream. Get it right, and we have mutually supportive, productive and interconnected systems. Get it wrong, and we have productive systems that continue to undermine their own success by damaging the health of nature, people and communities.

# Notes

## Chapter I  Landscapes Lost and Found

1  Data on food production analysed from the Food and Agriculture Organization's (FAO's) *FAOSTAT* database (www.fao.org). During the past 15 years, aggregate production in Europe has been largely stable due to supply management policies, whereas in the US it has grown by 35 per cent.

2  World population was 3.08 billion in 1960; 3.69 billion in 1970; 4.44 billion in 1980; and 5.27 billion in 1990. The annual growth rate of world population was 2.1 per cent in the late 1960s; had fallen to 1.3 per cent during the late 1990s; and is projected by the United Nations Population Fund, 1999, to fall further to 1 per cent by 2015, to 0.7 per cent by 2030, and to just 0.3 per cent by 2050.

3  For details of food policy analyses and challenges, see, in particular, materials from the International Food Policy Research Institute (IFPRI) (www. ifpri.org) and FAO (www.fao.org). For specific papers, see Pinstrup-Andersen and Cohen, 1999; FAO, 2000a; Pinstrup-Andersen et al, 1999; Delgado et al, 1999.

4  In the US, US$25 billion is spent each year by federal and state organizations to provide extra food for this 12 per cent of the national population who are food insecure. On diets and obesity, see Eisinger, 1998; Lang et al, 1999; WHO, 1998, 2001; FAO, 2000a, 2000b; Lang et al, 2001.

5  For more on the nutrition transition, see Popkin, 1998. During the period to 2020, the urban population in developing countries is expected to double to nearly 3.5 billion, while rural numbers will grow by only 300 million to 3 billion. The numbers of urban people will, for the first time in human history, have exceeded those in rural areas. Such a change will also affect food consumption. As rural people move to urban areas, and as urban people's disposable incomes increase, so they tend to go through the nutrition transition – particularly from rice to wheat, and from coarse grains to wheat and rice. They also tend to eat more livestock products, processed foods, and fruit and vegetables.

6  The annual demand for cereals is predicted by the IFPRI to grow from 1400 million tonnes in 1995 to 2120 million tonnes by 2020. Of this, 2.12 billion tonnes, 48 per cent will be for food and 21 per cent for animal feed in developing countries, and 8 per cent for food and 23 per cent for animal feed

in industrialized countries. Meat demand is expected to double by 2020 in developing countries to 190 million tonnes per year, and increase by one quarter in industrialized countries to 120 million tonnes.

7  On average, intensive livestock fed a diet of grain and silage produce only 1 *megajoule* of meat for every 3 megajoules of grain eaten. There is another problem. As we eat more meat, so cereals are increasingly diverted for livestock feed, and those in food poverty stay in poverty. Today, 72 per cent of all cereals consumed in industrialized countries is for livestock feed. In developing countries, the pattern is inverted, with 74 per cent of all cereal still being directly consumed by humans. On the livestock revolution, see Delgado et al, 1999; Rosegrant et al, 1997. See also White, 2000; Seidl, 2000.

8  I would like to acknowledge Linda Hasselstrom for her fine essay 'Addicted to Work' (1997), in which the idea of converting human history to generations is developed. My estimates are slightly different from hers, as I use the dates of 7 million years before present (BP) for human divergence from apes, 12,000 BP for the start of agriculture, and a figure of 20 years for the average generation length. For more, see Diamond, 1997.

9  For more on collective action in agriculture, and the effects of modern agriculture, see Balfour, 1943; Huxley, 1960; Palmer, 1976; Picardi and Siefert, 1976; Jodha, 1990; Ostrom, 1990; Bromley, 1992; Pretty, 1995, 1998; Berkes and Folke, 1998; Kothari et al, 1998. For a comprehensive review of how social systems have developed management practices based on ecological knowledge for dealing with the dynamics of local ecosystems, see Berkes and Folke (eds), 1998. Petr Kropotkin was one of the first writers to give collective action prominence in his 1898 book *Mutual Aid*. He drew attention to the '*immense importance which the mutual-support instincts, inherited by mankind for its extremely long evolution, play even now in our modern society, which is supposed to rest on the principle "every one for himself, and the State for all"*' (pxv). Kropotkin drew upon the history of guilds and unions in many countries, including craft guilds of mediaeval cities, brotherhood groups of Scandinavia, *artéls* and *druzhestva* of Russia, *amhari* of Georgia, communes of France, and the *Geburschaflen* of Germany. '*Organizations came into existence wherever a group of people – fishermen, hunters, travelling merchants, builders or settled craftsmen – came together for a common pursuit*' (p171).

10  Cato, M P (1979) in Hooper, W D (ed) (revised Ash, H B) *Di Agri Cultura*. Marcus Porcius Cato, *On Agriculture*, and Marcus Terentius Varro, *On Agriculture*, Harvard University Press, Cambridge, Massachusetts.

11  See Li Wenhua (2000), *Agro-Ecological Farming Systems in China*. For a classic text on Chinese agriculture, see F H King (1991) *Farmers of Forty Centuries*. Here he introduces the idea of permanent agriculture. For a review of history of innovation in China, see Temple (1986) *China: Land of Discovery and Invention*.

12  The dates of births and deaths for these four are: Francis Bacon (1561–1626), Galileo Galilei (1524–1642), René Descartes (1596–1650) and Isaac Newton (1642–1727). Though the Enlightenment provided the boost for modern science's disconnection, it is important to note that it has not affected all sciences in the same way. There is great diversity within scientific disciplines.

There is also great significance in the institutional location of, and pressures on, scientific research.

13 The voices of the Earth quotes are all from Senanayake in Posey, 1999, pp125–152.

14 These principles are common in Brazil and Ecuador, where Daniel Mataho Cabixi of the Paraci people and Cristina Gualinga of the Quicha talk about the fundamental connections and difficulties in modern times. First Cabixi:

> *We have our mythological hero who is called Wasari. . . Wasari allocated territory to the different Paraci groups. . . and taught them the technologies of hunting and preparing and consuming natural resources. Wasari further established political and economic principles revealing how to deal with other human beings and nature. . . Our traditional territory of 12 million hectares. . . has been reduced to 1200 hectares today. Now the Paraci have to face a number of serious limitations in order to survive.*

Says Gualinga: *'Nature, what you call biodiversity, is the primary thing that is in the jungle, in the river, everywhere. It is part of human life. Nature helps us to be free, but if we trouble it, nature becomes angry. All living things are equal parts of nature and we have to care for each other'* (in Posey, 1999).

15 For the story of the Cree, see Berkes, 1998.

16 See Hardin (1968) *The Tragedy of the Commons.*

17 See Olson, 1965.

18 For the study of 82 villages in India, see Jodha, 1990, 1991. For more on the effects of local institutions on natural resources, see Scoones, 1994; Pretty and Pimbert, 1995; Leach and Mearns, 1996; Pretty and Shah, 1997; Ghimire and Pimbert, 1997; Singh and Ballabh, 1997.

19 For the dangers of 'theming' our urban and rural spaces and the attempted manufacture of community, see Goin, 1992; Garreau, 1992; Barker, 1998.

20 For more on the different types of thinking about the environment and types of sustainability, including post-modern views, see Hutcheon, 1989; Naess, 1992; Worster, 1993; Benton, 1994; Soul and Lease, 1995; Rolston, 1997; Barrett and Grizzle, 1999; Dobson, 1999; Cooper, 2000a. It is interesting to note that the term 'ecology' has now come to mean much more than a scientific discipline describing natural processes; it is also a noun that defines the environment as a whole. For a good collection of the writings on the philosophies of environmentalism, see Sessions (1995) *Deep Ecology for the 21st Century.*

21 For more on the landscape scale, see Klijn and Vos, 2000; Foreman, 1997; Cooper, 2000a.

22 See Cronon et al, 1992; Deutsch, 1992; Brunkhorst et al, 1997.

23 See Benton, 1994; Gray, 1999, p63.

24 See Bennett, 1999, p104.

25 For Thoreau quote, see Nash, 1973, p84 – quoted, in turn, from a speech by Thoreau on 23 April 1851 to the Concord Lyceum. For Muir quote, see Oelschlaeger, 1991. See also Nash, 1973; Oelschlaeger, 1991; Schama, 1996. See also Vandergeest and DuPuis, 1996.

26   Worster, 1993, p5.

27   For a good review of Thoreau, see Oelschlaeger, 1991, pp133–171.

28   Quotes are from Thoreau's *A Winter Walk*, p167; and Thoreau's *Maine Woods*, pp93–95. Also, see *The Writings of H D Thoreau*, volumes 1-6, Princeton University Press, 1981–2000.

29   See Cooper, 2000b.

30   See Nash, 1973, p3. For a discussion of the static and dynamic nature of locality and our desire, or otherwise, to conserve it, see Scruton, 1998, in *Town and Country*. Common Ground, a UK charity, makes this point in the recent book on community orchards: *'defining beauty as mountains, and richness as rarity, has not only devalued the remainder, but it has diminished people's confidence to speak out for ordinary things. . . everyday places are as vulnerable as the special'.* In the commonplace and the everyday, we form deeper cultural relations with nature and the land. See Common Ground's *The Common Ground Book of Orchards*, 2000.

31   In Okri, 1996, *Birds of Heaven*, p26.

32   Rackham, 1986.

33   See Suzuki, 1999. For more on the stories of Japan, see Suzuki and Oiwa (1986) *The Japan We Never Knew*.

34   See Berry, 1998.

35   Mabey (1996) *Flora Britannica*. For quotes, see pp7, 162–163, 232.

36   Folklore author Ralph Whitlock (1979) suggests that *'all superstitions and customs are logical if looked at in the right way'.*

37   The contrast with India is striking. Darshan Shankar (1998) estimates that there are 1 million local people in India, such as traditional birth attendants, bone-setters, herbal healers and wandering monks, who still have extensive knowledge of the uses of plants and animals, including another 400,000 licensed practitioners of systems of medicine such as Ayurveda and Siddha. The number of plants used for medicinal, food, fodder and fuel uses can be extraordinary, with some 7500 species across India with described values. Individual groups may have knowledge of several hundred species, such as the Mahadev Koli tribals who use more than 600 species, and the Karjat tribals of the Western Ghats who use 509 species.

38   Of the 5000–7000 oral languages persisting worldwide, 32 per cent are in Asia, 30 per cent in Africa, 19 per cent in the Pacific, 15 per cent in the Americas, and 3 per cent in Europe. See Grimes, 1996.

39   Maffi, 1999, p31.

40   For more on the Tohono O'odham, see the work of Ofelia Zepeda at the University of Arizona on reinvigorating the language and its links to the land (www.u.arizona.edu/~mizuki/wain/wain0.html). The Tohono O'odham abandoned the more common term for their culture, Papago, in the 1980s, as it means 'bean eaters'.

41   Molina, 1998, p31.

42   Evers and Molina, 1987.

43   For a discussion of the use of language and rhetoric to describe the transformation of the western interior of the US, see Lewis (1988) in Cosgrove and Daniels (eds). For an analysis of the linkage between language and an

understanding of the land amongst the Innu of Canada, see Samson, 2002. Even when taught in school, young Innu are stripped of the experience of being on the land – they are taught a hunting language in a setting that presumes agriculture as the dominant activity rather than hunting.

44  See Cronon et al, 1992, pp4, 18.

45  On the work of Frederick Jackson Turner (1930), William Cronon and colleagues suggest that *'it would be a shame to lose the power of this insight just because Turner surrounded it with a lot of erroneous, misleading and wrong-headed baggage'* (p6).

46  In Nash, 1973, p23.

47  In Nash, 1973, p 27.

48  See Miles, 1992; Jennings, 1985.

49  See Cronon et al, 1992, p15.

50  Assuming a date of 30,000–35,000 years BP for human arrival on the Australian continent – see Diamond, 1997. See also Smith, 1985; Carter, 1987.

51  See Carter, 1987, pp9, 41, 54.

52  For more details on the problems of dryland salinity, see the Australian National Drylands Salinity Programme (www.lwrrdc.gov.au/ndsp/index.htm). See also Pannell, 2001, salinity policy.

53  See Moorehead, 1966; Smith, 1985, pix; Smith, 1987, p80.

54  Mark Twain is quoted in Miles, 1992, p52. For later quote, see pp65–66.

# Chapter 2 Monoscapes

1  Taking 1961 as the baseline of 100, total food production increased in the following 40 years (to the year 2000) to 245 for the whole world, to 252 for Africa, to 381 for Asia, to 296 for Latin America, to 202 for the US, to 168 for western Europe, to 570 for China, and to 155 for the UK. Again taking 1961 as a baseline of 100 for per-capita food production, the index in 2000 was 125 for the world, 91 for Africa, 176 for Asia, 128 for Latin America, 137 for the US, 142 for western Europe, 299 for China, and 139 for the UK. Data were analysed from the Food and Agricultural Organization's (FAO's) database, *FAOSTAT*, Rome.

2  Cosgrove and Daniels (1988) *The Iconography of Landscape*.

3  Barrell (1980) *The Dark Side of the Landscape*.

4  For a fine analysis of English landscape painting by Gainsborough, Richard Wilson, J M W Turner, John Constable and George Robert Lewis, see Prince, 1988, in Cosgrove and Daniels (eds).

5  Before 1714, there were eight parliamentary acts to enclose open fields. There were then 18 between 1714–1727 under George I; 229 between 1727–1760 under George II; and 2500 under George III and Queen Victoria between 1761–1844. For details of the enclosures of open fields and the commons, see Hoskins, 1955; Wordie, 1983; Mingay, 1977.

6  For more on the commons of England and Wales, see DETR, 1998; English Nature, 1999; Short and Winter, 1999; Short, 2000.

7  See Lord Ernle (Prothero, R) (1912) *English Farming. Past and Present* for a review of writers of the time, such as Taylor, S (1652) *Common Good*; Moore, A (1653) *Bread for the Poor. . . Promised by Enclosure of the Wastes and Common Grounds*; Lee J (1656) *A Vindication of a Regulated Enclosure*; Hartlib, S (1651) *Legacie*; Tusser, T (1580) *Five Hundred Points of Good Husbandry*; Fitzherbert, A (1523) *The Boke of Husbandrie*; Norden J (1607) *The Surveyors Dialogue*; Houghton J (1681) *A Collection of Letters for the Improvement of Husbandry and Trade*. For a comprehensive summary of 447 agricultural writers, see Donaldson, 1854.

8  See Ernle, 1912, p120.

9  See Cobbett, 1830.

10  See Pretty, 1991.

11  See Humphries, 1990.

12  From John Clare's journal entry for Wednesday, 29 September 1824, p37 in Tibble, A (ed) *The Journal; Essays; The Journey from Essex*, Carcanet New Press, Manchester. For a masterpiece on country life and social history of the time, see Clare's poem 'The Shepherd's Calendar' (ed Robinson et al). For more on his natural history writing, see Grainger (ed) 1993; Blythe, 1999. Clare's own village of Helpston in Northamptonshire was enclosed 1816, and throughout his writings in journals and poems he referred to the disappearance of secret pathways, narrow lanes, old stone pits and diverse hedgerows. These are the long-vanished features of rural life that form the almanac of 'The Shepherd's Calendar', from the opening in January '*withering and keen the winter comes, While comfort flies in close shut rooms*' to December's close with '*And boiling, elder berry wine, To drink the Christmas eve's good bye*'. The losses, though, continued after enclosure. In 1825, he says in his journal: '*I thought that fresh intrusions would interrupt and spoil my solitudes after the Inclosure they despoil a boggy place that is famous for orchids at Royce Wood end*' (Grainger, 1983, p169).

13  Quoted in Ernle, 1912, pp115–116.

14  For the Black Act, see E P Thompson (1975) *Whigs and Hunters*.

15  Thompson, 1975, pp91, 156.

16  For conflicts in Germany, France, Mexico and Russia, see Engels, 1956; Lewis, 1964; Bloch, 1978; Shanin, 1986. For more on the value of cooperation in rural communities throughout Europe, see also Blum, 1971.

17  Gadgil and Guha (1992) *This Fissured Land*, p2.

18  Jodha, 1988, 1990.

19  For the review of the value of the commons in India and Africa, see Beck and Naismith, 2001. See also Pasha, 1992; Beck, 1994; Agarwal, 1995; Iyengar and Shukla, 1999; Beck and Ghosh, 2000.

20  Quoted in Gadgil and Guha, 1992, pp88–89, 121.

21  For a good survey of the specific roles of women in the collective management of natural resources, see Agarwal, 1997. On the folly of privatizing and standardizing nature, and of letting only a few control power, see Steinberg, 1995.

22  For details of the Balinese *subaks* and effects of modern rice in Indonesia, see Collier et al, 1973; Poffenberger and Zurbuchen, 1980; Pretty, 1995a.

23 For more on the nature of *adat*, see Geertz, 1993. Also see Segundad. 1999 and Gapor, 2001.

24 See Peluso, 1996. She also makes the following statement: '*By depicting resources users (the local ones) as wild, destructive (or illiterate, uneducated, backward or non-innovative), state resource management agencies think they can justify their use of militaristic environmental protection.*'

25 See Pretty and Pimbert, 1995; Scott, 1998.

26 See Dipera, 1999, pp131–132.

27 Adams and McShane, 1992. For a history of African land use, see also Reader, 1997. Blaikie and Jeanrenaud (in Ghimire and Pimbert, 1997) quote Colonel Mervyn Cowin, an early preservationist who helped to set up the Serengeti National Park, which was to be a '*cultured person's playground*', with the main purpose to '*protect nature from the natives*'.

28 For details of the Barabaig, see Lane, 1990, 1993; Lane and Pretty, 1990. Partial evaluations were by Stone, 1982, and Young, 1983, both quoted in Lane, 1990. The elders are quoted by Paavo, 1989, and recorded by Lane, 1990.

29 See Gadgil and Guha, 1992, p125.

30 For an alternative view, that British foresters worked with a genuine interest in conservation rather than simply for maximizing economic returns, see Grove, 1990, in MacKenzie (ed) *Imperialism and the Natural World*. See also Arnold (1996) *Colonising Nature*, Chapter 6.

31 In Kothari et al, 1989.

32 See Roy and Jackson, 1993; Colchester, 1997; Duffy, 2000.

33 Gadgil, 1998, p229; Khare, 1998.

34 See Muir (1911) *A Summer in the Sierra*. For quotes used in this section, see pp31, 43–47, 77. For a comprehensive text of all of John Muir's eight journeys, see the single volume *The Eight Wilderness-Discovery Books (Story of My Boyhood and Youth; A Thousand Mile Walk in the Gulf; The First Summer in the Sierra; The Mountains of California; Our Natural Parks; The Yosemite; Travels in Alaska; Steep Trails*).

35 See Schama, 1996.

36 This enclave thinking is nowhere put better than by Avery, 1995, in his book *Saving the Planet with Pesticides and Plastic*.

37 The US Wilderness Act of 1964 defines wilderness as a place '*where man himself is a visitor who does not remain*', in Gómez-Pompa and Kaus, 1992.

38 See Nash, 1973.

39 The data on protected areas have been analysed from the comprehensive database at UNEP/WCMC (1997) *United Nations List of Protected Areas* (www.unep-wcmc.org/protected_areas/data/un-eaanalysis.htm).

40 The largest designators include the US and the UK (21 per cent); New Zealand (24 per cent); Germany (27 per cent); Tanzania (28 per cent); Australia (29 per cent); Denmark (32 per cent); Ecuador and Saudi Arabia (34 per cent); Kiribati (39 per cent); Belize (40 per cent); Venezuela (61 per cent); and Slovakia (76 per cent).

41 **Table 2.1** *Number and area of protected areas according to protection regimes (end of 1990s)*

|  | Africa | Asia and Pacific | Latin America and Caribbean | Rest of World | Total |
|---|---|---|---|---|---|
| *Number of protected areas* | | | | | |
| Total | 1254 | 3706 | 2362 | 23,028 | 30,350 |
| Number in categories 1–3 (nature reserves, wildernesses, national parks and monuments) | 346 | 944 | 936 | 8,478 | 10,704 |
| Number in categories 4–6 (habitat and species areas, protected landscapes, managed resources) | 908 | 2762 | 1426 | 14,550 | 19,646 |
| Proportion in categories 1–3 (%) | 28% | 25% | 40% | 37% | 35% |
| *Area of protected areas (million square kilometres)* | | | | | |
| Total area | 2.06 | 1.85 | 2.16 | 7.16 | 13.23 |
| Area in categories 1–3 (strict protection) | 1.21 | 0.72 | 1.37 | 3.82 | 7.12 |
| Area in categories 4–6 (managed resources) | 0.85 | 1.13 | 0.79 | 3.34 | 6.11 |
| Proportion in categories 1–3 (%) | 59% | 39% | 63% | 53% | 54% |

*Source:* adapted from UNEP-WCMC, 2001

42 In Peluso, 1996.

43 See Manning, 1989; Kothari et al, 1989; West and Brechin, 1992; Oelschlaeger, 1991; Pimbert and Pretty, 1995.

44 See Gómez-Pompa and Kaus, 1992. Also Bruner et al, 2001.

45 For a summary of recent debates on functionalist (those seeing humans as part of and embedded in nature) and compositional (those who only see humans as destroyers of pristine nature) positions, see recent articles in *Conservation Biology* by Callicott et al (1999), critical responses by Hunter (2000) and Willers (2000), and responses to these by Callicott et al (2000). For a view from the 'nay-sayers', those who disagree with community participation or any use of resources in protected areas, see Spinage, 1998.

46 For *satochi* and *satoyama*, see Environment Agency of Japan, 1999. *Satoyama* are hilly regions blessed with coppiced forests, natural spring water, and a stable farming environment with little damage from floods or dry spells. For the best review of Japanese art of the Edo period, see Royal Academy of Arts, 1981.

Many of the most famous images of Japan come from the Edo period, such as Hokusai's images of Mount Fuji painted in the 1830s. For more on the Japanese commons, see McKean, 1985, 1992. For more on cultures and land, see Suzuki and Oiwa, 1996. For a comprehensive review of Mansanobu Fukuoka's invention of natural farming, see Fukuoka (1985) *The Natural Way of Farming*.

47  See Newby, 1988. For a discussion of landscape ecology and the value of patchiness, see Selman, 1993. For a review of the value of diversity in agro-ecosystems and in landscapes, see Swift et al, 1996. For a review of mosaic landscapes, see Ryszkowski, 1995; Klijn and Vos, 2000.

48  The post-modern, according to architect Charles Jenks, has five elements. It works on several levels at once; it is a hybrid drawing on many traditions; it is rich in language, particularly metaphor; it is new and enduring; and it responds to the multiplicity of a particular place. All of these features apply directly to agricultural systems. For an excellent discussion of the authoritarian high-modernism of Le Corbusier, see Scott (1998) *Seeing Like a State*, Chapters 3–4, pp87–146.

49  In most industrialized countries, there is yet to emerge the idea that nature and food production can come from the same process. In farmers' and policy-makers' minds, these areas are separate. It is fine to 'green the edge' of farming, but not yet acceptable to 'green the middle'. In Australia, according to Ruth Beilin, this means that the last decade's extraordinary Landcare movement of rural social organization '*is doomed to act within the existing paradigm of productive landscapes, with conservation zones created on the edges*'. Government money furthers this process of keeping conservation outside of productive agriculture. In Europe, government programmes have supported agri-environmental programmes to create patches of wildlife and non-farmed habitat precisely in those areas that not highly productive. See Beilin, 2000, p5.

50  For quotes, see Thoreau's *Walden*, pp100, 164–180, 362.

51  See Lopez, 1986, pxxii.

52  Quoted in Arnold, 1996.

53  For a summary of recent landscape ecology and science, see Klijn and Vos, 2000.

54  For a discussion of the agricultural expansion in El Péten, see Katz, 2000.

## Chapter 3  Reality Cheques

1  On the cheapness of food, Donald Worster recognized this about a decade ago:

*The farm experts merely assume, on the basis of marketplace behaviour, that the public wants cheapness above all else. Cheapness, of course, is supposed to require abundance, and abundance is supposed to come from greater economies of scale, more concentrated economic organization, and more industrialized methods. The entire basis for that assumption collapses if the marketplace is a poor or imperfect reflector of what people want* (Worster, 1993, *The Wealth of Nature*, p87).

2  See Astor and Rowntree, 1945, pp33, 47.

3  For more on George Stapledon, see Conford (1988) *The Organic Tradition*, pp192–193, 196–197.

4  Despite my regular use of these five terms as capitals, I agree with the misgivings that many have. Capital implies an asset, and assets should be looked after, protected and built up. But as a term, capital is problematic for two reasons: it implies measurability and transferability. Because the value of something can be assigned a single monetary value, it appears to matter not if it is lost because we could simply allocate the required money to buy another, or transfer it from elsewhere. However, we know that this is nonsense. Nature and its cultural and social meanings are not so easily replaceable. Nature is not a commodity, reducible only to monetary values. Nonetheless, as terms, natural capital and social capital have their uses in reshaping thinking around basic questions, such as what is agriculture for, and what system works best? For further discussions, see Benton, 1998; Bourdieu, 1986; Coleman, 1988, 1990; Putnam, 1995; Costanza et al, 1999; Benton, 1998; Carney, 1998; Flora, 1998; Grootaert, 1998; Ostrom, 1998; Pretty, 1998; Scoones, 1998; Uphoff, 1998; Pretty and Ward, 2001.

5  Worster, 1993, p92.

6  See Pretty, 1998; FAO, 1999; Conway and Pretty, 1991; Altieri, 1995; Pingali and Roger, 1995; Conway, 1997.

7  For more on the definitions and principles of sustainable agriculture, see Altieri, 1995, 1999; Thrupp, 1996; Conway, 1997; Pretty, 1995, 1998; Drinkwater et al, 1998; Tilman, 1998; Hinchliffe et al, 1999; Zhu et al., 2000; Wolfe, 2000.

8  An externality is any action that affects the welfare of, or opportunities available to, an individual or group without direct payment or compensation, and may be positive or negative. See Baumol and Oates, 1988; Pearce and Turner, 1990; EEA, 1998; Brouwer, 1999; Pretty et al, 2000. Economists distinguish between 'technological' or physical externalities, and 'pecuniary', or price effect, externalities. Pecuniary externalities arise, for example, when individuals or firms purchase or sell large enough quantities of a good or service to affect price levels. The change in price levels affects people who are not directly involved in the original transactions, but who now face higher or lower prices as a result of those original transactions. These pecuniary externalities help some groups and hurt others, but they do not necessarily constitute a 'failure' of the market economy. An example of a pecuniary externality is the rising cost of housing for local people in rural villages that results from higher-income workers from metropolitan areas moving away from urban cores and bidding up the price of housing in those villages. Pecuniary externalities are a legitimate public concern, and may merit a public policy response. Technological externalities, however, do constitute a form of 'market failure'. Dumping pesticides sewage into a lake, without payment by the polluter to those who are adversely affected, is a classic example of a technological externality. The market fails in this instance because more pollution occurs than would be the case if the market or other institutions caused

the polluter to bear the full costs of its actions. It is technological externalities that are commonly termed 'externalities' in most environmental literature (see Davis and Kamien, 1972; Common, 1995; Knutson et al, 1998).

9 For more on the value of nature's goods and services, see Abramovitz, 1997; Costanza et al, 1997, 1999; Daily, 1997; and *Ecological Economics*, 1999, vol 25 (I).

10 See Pimentel et al, 1992, 1995; Rola and Pingali, 1993; Pingali and Roger, 1995; Evans, 1995; Steiner et al, 1995; Fleischer and Waibel, 1998; Waibel and Fleischer, 1998; Bailey et al, 1999; Norse et al, 2000. The data from these studies are not easily comparable in their original form as different frameworks and methods of assessment have been used. Methodological concerns have also been raised about some studies. Some have noted that several effects could not be assessed in monetary terms, while others have appeared to be more arbitrary (eg the US$2 billion cost of bird deaths in the US is arrived at by multiplying 67 million losses by US$30 a bird: see Pimentel et al, 1992). The Davison et al (1996) study on the Netherlands agriculture was even more arbitrary. It added an estimate of the costs that farmers would incur to reach stated policy objectives, and these were based on predicted yield reductions of 10–25 per cent arising from neither cheap nor preferable technologies, which led to a large overestimate of environmental damage (see Bowles and Webster, 1995; Crosson, 1995; Pearce and Tinch, 1998; van der Bijl and Bleumink, 1997).

11 On the effects of pesticides in rice, see Rola and Pingali, 1993; Pingali and Roger, 1995.

12 Hartridge and Pearce, 2001.

13 See Pretty et al, 2000, 2001. These are likely to be conservative estimates of the real costs. Some costs are known to be substantial underestimates, such as acute and chronic pesticide poisoning of humans, monitoring costs, eutrophication of reservoirs and the restoration of all hedgerow losses. Some currently cannot be calculated, such as dredging to maintain navigable water, flood defences, marine eutrophication and poisoning of domestic pets. The costs of returning the environment or human health to pristine conditions were not calculated, and treatment and prevention costs may be underestimates of how much people might be willing to pay in order to see positive externalities created. The data also do not account for time lags between the cause of a problem and its expression as a cost, as some processes that have long since ceased may still be causing costs. Some current practices, furthermore, may not yet have caused costs, and this study did not include the externalities arising from transporting food from farms to manufacturers, processors, retailers and, finally, to consumers.

14 See Pretty et al, 2001.

15 The government's Office of the Director General of Water Services sets industry price levels every five years, which determine the maximum levels of water bills and specify investments in water quality treatment. During the 1990s, the water industry undertook pesticide and nitrate removal schemes, resulting in the construction of 120 plants for pesticide removal and 30 for nitrate removal (Ofwat, 1998). Ofwat estimates that water companies will spend a further

UK£600 million between 2000–2005 on capital expenditure alone due to continuing deterioration of 'raw water' quality due to all factors. Ofwat predicts capital expenditure for pesticides to fall to UK£88 million per year at the end of the 1990s/early 2000s, and for nitrate to fall to UK£8.3 million per year.

Although Ofwat has sought to standardize reporting, individual companies report that water treatment costs in different ways. Most distinguish treatment for pesticides, nitrate, *Cryptosporidium*, and several metals (iron, manganese and lead). The remaining treatment costs for phosphorus, soil removal, arsenic and other metals appear under a category labelled 'other'. Of the 28 water companies in England and Wales, 3 report no expenditure on treatment whatsoever; and a further 3 do not disaggregate treatment costs, with all appearing under 'other'. Twenty companies report expenditure on removal of pesticides, 11 on nitrates, and 10 on *Cryptosporidium*. It is impossible to tell from the records whether a stated zero expenditure is actually zero, or whether this has been placed in the 'other' category. Using Ofwat and water companies' returns, we estimate that 50 per cent of expenditure under the 'other' category refers to the removal of agriculturally related materials.

16   We originally calculated the annual external costs of these gases to be UK£280 million for methane, UK£738 million for nitrous oxide, UK£47 million for carbon dioxide, and UK£48 million for ammonia. But a later analysis of marginal costs (Hartridge and Pearce, 2001) suggests that costs are lower for methane (UK£83 million), nitrous oxide (UK£290 million) and carbon dioxide (UK£22 million), putting the total at UK£444 million per year.

17   DETR, 1998a, 1998b; Pretty, 1998; Campbell et al, 1997; Pain and Pienkowski, 1997; Mason, 1998; Siriwardena et al, 1998; Krebs et al, 1999.

18   Repetto and Baliga, 1996; Pearce and Tinch, 1998; HSE, 1998a; 1998b; Pretty, 1998.

19   Fatalities from pesticides at work in Europe and North America are rare: one a decade in the UK, and eight a decade in California. In the UK, a variety of institutions collect mortality and morbidity data; but in California, where there is the most comprehensive reporting system in the world, official records show that 1200–2000 farmers, farm workers and the general public are poisoned each year (see CDFA, passim; Pretty, 1998). There appears to be greater risk from pesticides in the home and garden where children are most likely to suffer. In Britain, 600–1000 people need hospital treatment each year from home poisoning.

20   On food poisoning in the UK, see PHL, 1999; Evans et al, 1998; Wall et al, 1996. For a study of food-borne illnesses in Sweden, see Lindqvist et al, 2001. When bovine spongiform encephalopathy (BSE) was first identified in late 1986, research confirmed that it was a member of a group of transmissible diseases occurring in animals and humans. It appeared simultaneously in several places in the UK, and has since occurred in native-born cattle in other countries. By mid 2001, more than 180,000 cases had been confirmed in the UK, the epidemic having reached a peak in 1992. The link between BSE and variant Creutzveldt-Jakob Disease (CJD) in humans was confirmed in 1996; 100 deaths

from CJD have occurred to 2001. The annual external costs of BSE were UK£600 million at the end of the 1990s. See NAO, 1998; WHO, 2001. By mid 2001, there had been 181,000 cases of BSE reported in the UK, 648 in Ireland, 564 in Portugal, 381 in Switzerland, 323 in France, 81 in Germany, 46 in Spain, and 34 in Belgium. For more on the implications of BSE and lessons to be learned, see Lobstein et al, 2001; Millstone and van Zwanenberg, 2001.

21 For an excellent review of food crises and the need for new thinking in food systems, see Lang et al, 2001. See also Waltner-Toews and Lang, 2000.

22 Donald Worster (1993, p18) points out that this was not, of course, the end of the story. Control through the levees did not stop conflicts between farmers who wanted water for irrigation, and others who wanted to protect the natural habitat of waterfowl. The levees also did not stop pesticides and nutrients from washing off the fields, or stop the emergence of livestock feedlots, with their massive production of animal wastes.

23 On the effects of changes in the German landscape from flooding, see van der Ploeg et al, 1999, 2000. Vo-Tong Xuan, rector of Angiang University in Vietnam, notes similar problems in the Mekong Delta, where farmers have switched from one crop of floating rice per year to three short duration crops of modern varieties, which has led to the occurrence of floods on an annual basis.

24 On the effects of water control in the Japanese landscape, in particular on paddy rice fields, see Minami et al, 1998; Kato et al, 1997; OECD, 2000.

25 On the externalities of Chinese agriculture, see Norse et al, 2000.

26 FAO, 2000a.

27 On the values of wetlands, see Heimlich et al, 1998. For a study on eutrophication costs, see Pretty et al ( 2002) *An Assessment of the Costs of Eutrophication*. See also Postel and Carpenter (1997) in Daily (ed) *Nature's Services*; Ewel (1997) in Daily (ed) *Nature's Services*. For a study showing that the costs of creating wetlands are less than for constructing treatment plants, see Gren, 1995.

28 See Keeny and Muller, 2000.

29 We distinguished between value–loss costs arising from the reduced value of clean or non-eutrophic (nutrient-enriched) water, and the direct costs incurred in responding to eutrophication, plus the costs of changing behaviour and practices in order to meet legal obligations. Value–loss costs, by definition, represent a loss of existing value, rather than an increase in costs, and are divided into two categories: use values and non-use values. Use values are associated with private benefits gained from actual use (or consumption) of ecosystem services, and can include private-sector uses (eg agriculture, industry), recreation benefits (eg fishing, water sports, bird watching), education benefits, and general amenity benefits. Non-use values are of three types: option values, bequest values and existence values. See Pretty et al, 2001; also Mason, 1996; Environment Agency, 1998.

30 Total fertilizer consumption (nitrogen, phosphate and potassium) for the world was 138 million tonnes (mt) in the year 2000, comprising 83 mt of nitrogen, 32 mt of phosphate, and 22 mt of potassium. Nitrogen consumption in western Europe was 17 mt, in North America 21 mt, in South Asia 21 mt,

in the Russian states 38 mt, and in China 38 mt. World consumption of all fertilizer has grown from 30 mt in 1960 (when nitrogen consumption was 11 mt, phosphate 11 mt and potassium 8 mt). Data are from the International Fertilizer Industry Association, Paris.

31  WHO (1998) *Obesity: Preventing and Managing the Global Epidemic.* See also Oster et al, 1999.

32  See WHO (2001) *Food and Health in Europe: A Basis for Action.*

33  For details of food-borne illnesses, see CDC, 2001; Kaeferstein et al, 1997; Mead et al, 1999. For US Department of Agriculture (USDA) data on microbial infections in farm animals, see the USDA website (www.usda. fsis.usda.gov). For costs of antibiotic resistance, see the National Institute of Allergy and Infectious Diseases. For Centres for Disease Control (CDC), see the CDC website (www.cdc.gov).

34  See Buzby and Robert, 1997; WHO (2001) *Food and Health in Europe.*

35  **Table 3.1**  *Incidence of microbial infection in US farm animals*

|  | Proportion of individuals with infectious bacteria (%) | | | |
|  | Broiler chickens | Turkeys | Pigs | Cattle |
| --- | --- | --- | --- | --- |
| *Clostridium* | 43 | 29 | 10 | 8 |
| *Campylobacter* | 88 | 90 | 32 | 1 |
| *Salmonella* | 29 | 19 | 8 | 3 |
| *Staphylococcus* | 65 | 65 | 16 | 8 |

*Source*: USDA data

36  For more on antibiotics and the emergence of resistance, see Harrison and Lederberg, 1998; Wise et al, 1998; House of Lords, 1998.

37  See Havelaar et al, 2000; WHO, 2001; FAO, 2001.

38  See Willis et al, 1993; Foster et al, 1997; Stewart et al, 1997; Hanley et al, 1998.

39  See Cobb et al, 1998.

40  Data are from the Countryside Agency and English Tourism Council surveys: 1968 million tourist-days were spent in the UK in 1998, of which 433 million were to the countryside, 118 million to the seaside, and 1299 million to towns.

41  IATP, 1997.

42  Watershed Agricultural Council. Catskill/Delaware Watershed Complex (www.iatp.org/watersheds).

43  Pretty and Ball, 2001.

44  See Pretty and Ball, 2001; Swingland et al, 2002.

45  Growing empirical evidence on the costs of compliance with environmental regulations and taxes suggests that there has been little or no impact on the overall competitiveness of businesses or countries, with some indications that they have increased efficiency and employment. See EEA, 1996; 1999; Smith

and Piacentino, 1996; Ekins, 1999; OECD, 1997; Jarass and Obermair, 1997; Rayement et al, 1998; DETR, 1999a; Ribaudo et al, 1999.

46  See Ekins, 1999, for a comprehensive review of environmental taxes.

47  See Myers, 1998; Potter, 1998; Dumke and Dobbs, 1999; Hanley and Oglethorpe, 1999.

48  For more on Cuba, see Rossett, 1997, 1998; Funes, 2001.

49  Swiss Agency for Environment, Forests and Landscape and Federal Office of Agriculture, 1999, 2000. See also Dubois et al, 2000.

# Chapter 4  Food for All

1  See Maxwell and Frankenberger, 1992; Hoddinott, 1999; ACC/SCN, 2000; Smith and Haddad, 1999.

2  The effect of dietary improvements can be dramatic, and nutritionists have long considered the positive effects of supplementing diets. The effect of the treating of Indonesian children with vitamin A tablets has been shown to reduce child mortality by 30 per cent (Smith and Haddad, 1999). Other micronutrients, such as vitamins B, D, folic acid and iron, could be added to wheat flour; but rice is more difficult. It cannot easily be used as a fortification vehicle, and poor people often cannot get access to sufficient quality and quantity of foods that are rich in micronutrients, vitamins and minerals.

3  See ACC/SCN, 2000.

4  See Chapter 3 of this book for more on the real costs of food production. Key references are Balfour, 1943; Carson, 1963; Conway and Pretty, 1991; Pimentel et al, 1992, 1995; Steiner et al, 1995; EEA, 1998; Waibel and Fleischer, 1998; Norse et al, 2000; Pretty et al, 2000a, 2000b; McNeely and Scherr, 2001; Uphoff, 2002.

5  For a selection of books on sustainable agriculture, see Altieri, 1995; Thrupp, 1996; Conway, 1997; Pretty, 1995a, 1998; Hinchliffe et al, 1999; NRC, 2000; Uphoff, 2002.

6  We used a four-page questionnaire as the main survey instrument for sustainable agriculture projects and initiatives. It addressed: key impacts on total food production, and on natural, social and human capital; the project/initiative structure and institutions; details of the context and reasons for success; and spread and scaling-up (institutional, technical and policy constraints). The questionnaire was centred on an assets-based model of agricultural systems, and was developed in order to understand both the role of these assets as inputs to agriculture and the consequences of agriculture upon them. The questions were also formulated with regard to the nine types of sustainable agriculture improvement identified as the conceptual base for this project. We collated all returned questionnaires and secondary material, and added this to the country databases. All data sets were re-examined in order to identify gaps and ambiguities, and correspondents were contacted again to help fill these. We established trustworthiness checks by engaging in regular personal dialogue with respondents,

through checks with secondary data, and by critical review by external reviewers and experts. We rejected cases from the database on several grounds:

- where there was no obvious sustainable agriculture link;
- where participation was for direct material incentives (there are doubts that ensuing improvements persist after such incentives end);
- where there was heavy or sole reliance on fossil-fuel derived inputs for improvement, or on their targeted use alone (this is not necessarily to negate these projects, but to indicate that they are not the focus of this research);
- where the data provided in the questionnaire have been too weak; and
- where findings were unsubstantiated by the verification process.

However, we undoubtedly missed many novel, interesting and globally relevant projects/initiatives. We therefore present conservative estimates of what has been achieved, over what area, and by how many farmers. See Pretty and Hine, 2001.

7  The largest country representations in the 208 project dataset are India (23 projects/initiatives); Uganda (20); Kenya (17); Tanzania (10); China (8); the Philippines (7); Malawi (6); Honduras, Peru, Brazil, Mexico, Burkina Faso and Ethiopia (5); and Bangladesh (4). The projects and initiatives range very widely in scale – from 10 households on 5 hectares in one project in Chile to 200,000 farmers on 10.5 million hectares in southern Brazil.

8  The total arable land comprises some 1600 million hectares in 1995/ 1997, of which 388 million hectares are in industrialized countries, 267 million hectares in transition countries, and 960 million hectares in developing countries (see FAO, 2000a).

9  We found some 8.64 million small farmers practising sustainable farming on 8.33 million hectares, and 349,000 larger farmers in Argentina, Brazil and Paraguay farming with zero-tillage methods on 21 million hectares.

10

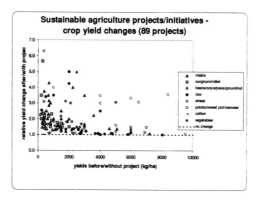

Source: Pretty and Hine, 2001; Pretty et al, 2002

**Figure 4.1** *Sustainable Agriculture Projects/Initiatives – Crop-Yield Changes (89 Projects)*

11 For a discussion on the value of ecosystem services in the soil, and the problems of agricultural intensification, see Daily et al, 1997. See also Cleaver and Schreiber, 1995; RCEP, 1996; World Bank/FAO, 1996; Hinchcliffe et al, 1999; Petersen et al, 2000; FiBL, 2000. For a summary of issues relating to sustainable management of land and soils, see FAO (1999) *The Future of Our Land.*

12 In Argentina, the amount of organic matter in soils fell over a century to 1990 from 5.5 per cent to 2.25 per cent; where no till systems are being used, it is increasing by 0.1 per cent per year (Roberto Peiretti, personal communication).

13 For a summary of the extent of soil degradation and its effects, see Koohafkan and Stewart, 2001; Smaling et al, 1997; Henao and Baanante, 1999.

14 See Reicosky, 1997; Sanchez et al, 1997; Sorrenson et al, 1998; de Freitas, 2000; Bunch, 2000. Some plants are called phosphate releasing, not because they release phosphate but because acids secreted from roots solubilize iron and aluminium phosphates in the soil. See Sanchez and Jama, 2000.

15 For effects of zero-tillage in Latin America, see Sorrenson et al, 1998; Petersen, 1999; de Freitas, 2000; Peiretti, 2000; Landers, 1999.

16 See Landers et al (2001) 'Environmental Benefits of Zero-Tillage in Brazil'.

17 Using 20 species of cover crops and green manures, Paolo Petersen and colleagues at AS-PTA have shown how small farmers can adopt zero-tillage systems without herbicides. See Petersen et al, 2000; von der Weid, 2000.

18 For details of the work of Rodale in Senegal, see Diop, 2000.

19 See Hamilton, 1998; Association for Better Husbandry, 2001.

20 One of the best examples of how changed rotations can transform whole agricultural systems comes from western Kenya, where farmers are using improved fallows using sesbania (*Sesbania sesban*), tephrosia (*Tephrosia vogelii*) and various species of *Crotalaria*. These are interplanted during the main rainy season maize crop, and capture 100–200 kilogrammes of nitrogen per hectare per year, with the added benefit of no transport costs and associated benefits for fuelwood production. Pedro Sanchez indicates that *'many farmers say hunger is now a thing of the past'*. See Sanchez et al, 1999; Sanchez and Jama, 2000.

21 Reij, 1996.

22 See Uphoff, 1999, 2000.

23 See *BAA Annual Report,* 2000. The eight companies in 2001 that control 90 per cent of the world pesticide market are Aventis, Syngenta, Monsanto, Du Pont, Bayer, Cyanamid, Dow and BASF. Data on the weight of pesticides used in agriculture are very difficult to access, not least because of the great differences in toxicity of active ingredients, with some products being high volume and low toxicity, and others low volume and high toxicity. According to the World Health Organization, some 3.1 billion kilogrammes were exported in 1985 at a value of US$16 billion. Since today's market is estimated to be in excess of US$31 billion, the total pesticides applied are likely to exceed 5 billion kilogrammes.

24 For more on farmer field schools in Asia, see Evelleens et al, 1996; Heong et al, 1999; Desilles, 1999; Jones, 1999.

25   The INTERFISH, NOPEST and GOLDA integrated pest management for rice and aquaculture projects in Bangladesh are supported by the UK Department for International Development (DFID) and the European Union, and are implemented by CARE.

26   For the semiochemicals research and outcomes, see Pickett, 1999; Khan et al, 2000.

27   The push–pull strategy involves trapping pests on highly susceptible trap plants (pull) and driving them away from the crop using a repellent intercrop (push). The forage grasses *Pennisetum purpureum* (Napier grass) and *Sorghum vulgare sudanense* (Sudan grass) attract greater ovi-position by stem borers (*Chilo spp*) than in cultivated maize. The non-host forage plants *Melinis minutiflora* (molasses grass) and *Desmodium uncinatum* (silver leaf) repel female stalk borers. Intercropping with molasses and sudan grass increases parasitism, particularly by the larval parasitoid *Cotesia sesamiae* and the pupal parasitoid *Dentichasmis busseolae*. *Melinis* contains several physiologically active compounds. Two of these inhibit ovi-position (egg-laying) in *Chilo*, even at low concentrations. Molasses grass also emits a chemical, (*E*)-4,8-dimethyl-1,3,7-nonatriene, which summons the borers' natural enemies. Napier grass also has its own defence mechanism against stem borers: when the larvae enter the stem, the plant produces a gum-like substance that kills the pest. And finally, intercropping maize with the fodder legumes *Desmodium uncinatum* (silver leaf) and *D. intortum* (green leaf) reduces infestation by the parasitic weed *Striga hermonthica* by a factor of 40 compared to maize monocrop.

28   For more on ecoagriculture, see McNeeley and Scherr, 2001.

29   For the effects of sustainable agriculture in Cuba, see Murphy, 1999; Funes, 2001; Kovaleski, 1999; Socorro Castro, 2001; Funes et al, 2002.

30   On aquaculture, see Brummet, 2000.

31   For reviews of the emergence of SRI in Madagascar, and its spread elsewhere, see the work of Norman Uphoff at Cornell University: Uphoff, 1999, 2001.

32   The farmed shrimps are *Penaeus* spp. or *Macrobrachium* spp.

33   See Li Wenhua, 2001.

# Chapter 5   Only Reconnect

1   For more on Suffolk Punches, see Suffolk Horse Society, Woodbridge, Suffolk (www.suffolkhorsesociety.org.uk). See also Evans (1960) *The Horse and the Furrow*. It is interesting to note that breeders of Suffolk Punches were also involved in breeding other now rare Suffolk animals, such as Red Poll cattle, Suffolk sheep and Large Black pigs.

2   The loss of horses from the landscape was not only due to their replacement with efficient machines. Horses regularly themselves suffered from grass sickness, a disease even today not fully understood, and their numbers had not recovered after the huge losses in the World War I. Reliable machinery thus helped to replace relatively unreliable horses.

3  After the progress made by Sue Heisswolf and Kevin Niemeyer with the Brassica Improvement Group, Brad Scholtz and colleagues (1998) found that the more pesticides applied to maize in Queensland, the lower the yields; and the less spray, the higher the yields. This echoes earlier research by Peter Kenmore and colleagues in Asia in the 1980s, who found that pest attack in rice was directly proportional to the amount of pesticides applied – pesticides killed the beneficial insects that were exerting good control of pests.

4

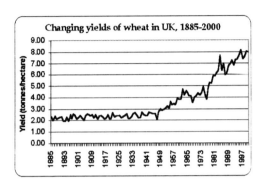

Source: DEFRA statistics

**Figure 5.1** *Changing Yields of Wheat in the UK, 1885–2000*

5  See Fuglie et al, 2000; USDA, 2001a

**Table 5.1** *Measures of Increasing Productivity in US Livestock Production*

|                              | 1955 | 1995 | % increase |
|------------------------------|------|------|------------|
| Beef cattle (kg beef/cow)    | 267  | 327  | 23         |
| Pigs (kg pork/sow)           | 357  | 680  | 90         |
| Dairy (kg milk/cow)          | 2643 | 7444 | 182        |
| Broilers (kg/bird)           | 1.39 | 2.11 | 52         |
| Layers (eggs/layer/year)     | 192  | 253  | 32         |

*Source:* Fuglie et al, 2000

6  In the US hog industry, just 2700 farms now have half of all American pigs. The other half are found on 140,000 farms, down from 900,000 since 1970. In the egg industry, 95 per cent of all 270 million layers are managed by 300 egg-producing operations, each owning flocks of 75,000 or more. Just ten companies control nine-tenths of all poultry production. Similar polarizing trends are evident in the European Union, where 6 per cent of farmers produce 60 per cent of cereals, and 15 rear 40 per cent of all farm animals. In the UK, just 52 pig holdings rear 80,000 pigs, while another 40,000 holdings raise 170,000 pigs in herds of less than 20. In the broiler chicken sector, 330 holdings raise 67 million broiler chickens (66 per cent of the total), while 720 holdings

have 59,000 birds in flocks of less than 1000. The picture is similar for laying chickens, with 300 holdings having 29 million laying chickens (80 per cent of the total), all in flocks of more than 20,000; yet 45 per cent of all holdings with layers have 400,000 birds in flocks of less than 100 birds. On arable farms, 8300 holdings have half of all the cereal area, while 32,000 have only a tenth of the area, all on farms of less than 20 hectares. Data are from the Department of the Environment, Farming and Rural Affairs (DEFRA; formerly MAFF) annual data, Economic and Statistics Group (www.defra.gov.uk/esg). For a good review of the future of the pig industry, see Harrington, 2000.

**Table 5.2** *Concentration of Operations in the UK*

| Sector | Large-scale operations | Small-farm sector |
|---|---|---|
| Cereal | 8300 holdings have 48% of cereal area (all on farms of greater than 100 hectares) | 31,000 holdings have 9% of the area (on farms of less than 20 hectares) |
| Laying chickens | 300 holdings have 29 million laying chickens (79% of the total), all in flocks of more than 20,000 | 45% of all holdings with layers, some 23,200, have 0.4 million birds in flocks of less than 100 birds |
| Broiler chickens | 334 holdings raise 67 million broiler chickens (66% of the total) | 722 holdings have 59,000 birds in flocks of less than 1000 (0.1% of the total) |
| Sheep (England and Wales) | 9700 holdings have 57% of total sheep in herds of more than 1000 | 18,000 holdings have 2.2% in herds of less than 100 |
| Beef cattle | 1300 holdings with 19% of national herd in herds of more than 100 | 30,000 holdings with 31% in herds of less than 30 |
| Pigs | 52 holdings have 80,000 pigs (13% of total) in herds of more than 1000 | 41,200 holdings have 170,000 pigs (30% of total) in herds of less than 20 |
| Dairy cattle | 922 holdings have 247,000 cattle (12% of herd) in herds of more than 200 | 5300 holdings have 69,000 cattle (35% of total) in herds of less than 30. |

*Source: MAFF, June 1999 Census data* (Economic and Statistics Group, www.defra.gov.uk/esg)

7  See Heffernan et al, 1999; Weida, 2000; Wesselink, 2001. In the dairy industry, the greatest gains in market share in recent decades have occurred in non-traditional milk-producing regions, such as California, Washington, Arizona and New Mexico, which now produce a quarter of all US milk. The traditional dairy-producing areas have suffered most – yet these tend to have smaller herd sizes and more diversified operations that also grow most of their own food. In Wisconsin, New York, Pennsylvania, Minnesota and Michigan, 40–70 per cent

of cattle are in herds of less than 100. By contrast, 96 per cent of all cattle in New Mexico, 78 per cent in California and 47 per cent in Washington are in herds of more than 500 animals. Now, only ten businesses account for half of all US milk production, a staggering 36 billion kilogrammes per year, and 50 account for three-quarters of total production.

8 **Table 5.3** Concentration Ratios in the US Food Chain, 1999

| Sector | Concentration ratio for top four firms (CR4) (%) | Notes on changes over time |
|---|---|---|
| Beef packers | 79 | Up from 72% in 1990 |
| Pork packers | 57 | Up from 37% in 1987 |
| Broiler producers | 49 | Up from 35% in 1986; top company produces 70 million kilogrammes per week |
| Turkey producers | 42 | Up from 31% in 1988 |
| Flour milling | 62 | Up from 44% in 1987 |
| Dry corn milling | 57 | – |
| Wet corn milling | 74 | Up from 63% in 1977 |
| Soybean crushing | 80 | Up from 54% in 1977 |
| Seed corn market | 69 | – |

Source: Heffernan, 1999

The same names keep reoccurring. ConAgra, for example, turns up at every stage of the food chain except for pesticide and machinery manufacture. ConAgra also owns about 1000 grain elevators, 1000 barges and 2000 railway cars. Cargill is in the top four firms which produce animal feed, rear cattle and process cattle. On the product side, 60–90 per cent of all wheat, maize and rice is marketed by only six transnational companies. One of these, Cargill, earns more from its coffee sales alone than the total income of any of the African countries from which it buys coffee. Again, is not all this efficiency for the best? Should we not be celebrating such advanced methods of producing more meat, milk and eggs from each animal and from each square metre of farm?

9 FAO/UNEP, 2000 (www.fao.org/dad-is). See also Blench, 2001. For more on domestic animals, see Domestic Animal Diversity Information System (DADIS) at www.dad.fao.org/cgi-dad/$cgi_dad.exe/summaries. Livestock experts consider that only when there are 100,000 individuals of a given species is a population stable and able to reproduce without genetic loss. Less than 10,000, and population numbers will decrease rapidly; below 1000, and the whole population is endangered, with size too small to prevent genetic loss. Europe has one quarter of the world's cattle, sheep, pig and duck breeds, and one half of horse, chicken and geese breeds. But in the five years to 1999, the number of mammalian breeds at risk grew from 33–49 per cent, and bird breeds at risk rose from 65–76 per cent.

**Table 5.4** *Number of Animal Breeds and Proportion at Risk of Extinction*

| Location | Number of breeds | Proportion at risk (%) |
|---|---|---|
| Europe | 2576 | 50 |
| North America | 259 | 35 |
| Asia and Pacific | 1251 | 10 |
| Sub-Saharan Africa | 738 | 15 |

Source: FAO, Rome

10  For a comprehensive review of global agroecosystems, see Wood et al, 2000. See also Rosset, 1999.

11  Aldo Leopold is quoted in Cooper, 1996.

12  USDA, 2001a (farm size and numbers data at www.usda.gov).

13  Steinbeck (1939) *The Grapes of Wrath.*

14  USDA, 2001b (farm statistics at www.ers.usda.gov/statefacts).

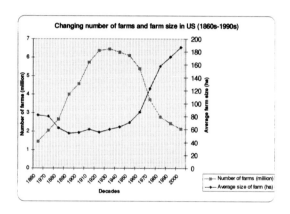

**Figure 5.2**  *Changing Number of Farms and Farm Size in US (1860s–1990s)*

15  See USDA (1998) *A Time to Act. National Commission on Small Farms.*

16  Professor Thomas Dobbs's evidence was given to the National Commission on Small Farms on 22 August 1997.

17  See Peterson, 1997.

18  See Goldschmidt, 1978 (1946); Perelman, 1976; Small Farm Viability Project, 1977. For a review of the pros and cons of the Goldschmidt hypothesis, see Lobao et al, 1993. See also Durrenberger and Thu, 1996.

19  See Lobao, 1990.

20  See Berry, 1977.

21  One outcome of the growing centralization of the food chain is the increase in unnecessary movements of food, both within and between countries. In the US, it has been estimated that each item of food travels 2000 kilometres from field to plate, causing damage to the environment through fossil fuel

emissions during transport and greater congestion on the roads. There are also many unnecessary food swaps between countries, with large amounts of the same products being imported and exported to and from the same countries. The UK, for example, exports 213,000 tonnes of pig meat each year, yet also imports 272,000 tonnes, resulting in a large number of unnecessary road movements (see Table 5.5).

**Table 5.5** *The UK's Food Swap To and From the European Union's 14 Member Countries and the Rest of the World, 2000*

| Sector | Domestic production (thousand tonnes) | Exports (thousand tonnes) | Imports (thousand tonnes) |
|---|---|---|---|
| Poultry | 1514 | 170 | 363 |
| Pigs | 738 | 213 | 272 |
| Cattle/calves | 706 | 9 | 202 |
| Sheep/lambs | 390 | 125 | 129 |
| Milk (million litres) | 14,054 | 423 | 124 |
| Wheat | 16,700 | 3505 | 930 |
| Barley | 6490 | 1730 | 51 |

*Source:* DEFRA, *Annual Statistics*, 2001

22  See Garkovich et al, 1995.
23  Butala, 2000.
24  See Coop and Brunkhorst, 1999; Swift et al, 1996.
25  See Kline, 1996.
26  See Butler-Flora and Flora, 1996.
27  See Tall, 1996.
28  Drennan Watson (personal communication, 2001) makes an interesting observation about Gaelic communities in Scotland: *'You don't ask a person in Gaelic where they come from, you ask them where they belong to. There is a special word for this belonging to a place – Duthcas – for which there is no equivalent in the English language.'*

29 **Table 5.6** *Proportion of the Food Pound Returned to Farmers*

|  | UK | US |
|---|---|---|
| Expenditure on food by consumers | UK£92.3 billion | US$788 billion |
| Farmers' gross receipts | UK£14.1 billion | US$208.7 billion |
| Farmers' expenditure on seeds, feedstuffs, pesticides, fertilizers, machinery fuel and insurance | UK£7.4 billion | US$181 billion |
| Farmers' net share of the food pound | 7.3% | 3.5% |

*Sources:* DEFRA and USDA statistics

30  See Bignall and McCracken, 1996.
31  There are many variations in interpretation in what organic farming should be doing. Some indicate that produce should only be marketed locally;

others are content for produce to travel long distances if consumers express a demand in the market.

32  For details of organic production, see USDA (2000) *Factbook* (www. usda.gov). See also Greene and Dobbs, 2001. The 3.3 million hectares of organic farming in the European Union in the year 2000 includes 420,000 hectares in the UK, 29,000 hectares in the Netherlands, 320,000 hectares in France, 455,000 hectares in Germany, 360,000 hectares in Austria, and 255,000 hectares in Italy. See Soil Association (2000) *The Organic Food and Farming Report*. See also Lampkin and Padel, 1994; Lampkin and Midmore, 2000.

33  See Balfour (1943) *The Living Soil*, p173. See also Conford (ed), 1988.

34  For a recent comparison of UK farming systems, see Tinker (2000) *Shades of Green*.

35  Nick Robins and Andrew Simms (2000) analysed a National Opinion Poll (NOP) survey conducted for Satish Kumar's journal *Resurgence*. Interestingly, when individuals were asked what they would do if they had a free day off, 38 per cent said they would spend time with friends and family, and 28 per cent said they would go for a walk in the country. Only 16 per cent said shopping, and 2 per cent said they would watch television. People were also asked how they'd like to be remembered – 68 per cent said as a good parent or kind person. Only 2 per cent said that they would like to be remembered as a wealthy or successful business person. Willingness to act for the environment was also high, with 24 per cent saying public protest was the best way to protect the environment, and 32 per cent saying public boycotts. Only 15 per cent said nothing could be done. This survey indicates considerable hope, with values of connectedness between people and with the environment far exceeding the consumerist, modernist mythology. This opposes some of the images of our modern world.

36  Data for visits and expenditure in the UK countryside come from the Countryside Agency (2001) and English Tourist Council (2000), who use the UK Leisure Day Visits Survey and UK tourism surveys to calculate the number of visits made to the countryside for leisure and recreational activities. In 1998, some 1.261 billion tourist day-visits were made, of which 72 per cent were to towns, 6 per cent to the seaside, and 22 per cent to the countryside. In addition to day visits, a further 172 million tourist trips are taken by UK and overseas residents, in which one or more nights are spent away, totalling 707 million days. Thus, there were 433 million visit-days to the countryside. Average spend per day/night is UK£16.90 for UK day visitors, UK£33.00 for UK overnight visitors, and UK£58.40 for overseas overnight visitors, putting the total spend at UK£11.02 billion per year.

37  The oldest environmental or countryside group in the UK is the Open Spaces Society, which was established in 1865 and was set up to protect commons in metropolitan areas. Most organizations aim to protect something perceived as threatened, such as birds and wildlife (eg the Royal Society for the Protection of Birds, the Sierra Club and the Wildlife Trusts); animal welfare (eg the Royal Society for the Prevention of Cruelty to Animals, with 25,000 members); the preservation of houses and properties (eg the National Trust);

the preservation of wildernesses (eg the Wilderness Society); the livelihood interests of specific rural groups (eg the National Farmers' Union, with 100,000 members; the Country Land and Business Association); the general milieu of the countryside (eg the Council for the Preservation of Rural England, with 45,000 members); access to the countryside (eg the Ramblers Association, with 112,000 members); and, more recently, the wider environment (eg Friends of the Earth and Greenpeace); or the very specific interests of hunting and shooting lobbies (eg the Moorland Association and the Countryside Alliance); or of protest movements against road-building and genetically modified crops (eg EarthFirst!). It is very difficult to say how many different people are members of these organizations, as it is likely that there many individuals who are members of several organizations. In addition, some distinguish between members and supporters, and affiliated organizations. It is also in the interests of some organizations to inflate their membership numbers in order to achieve more political recognition.

38  See Kloppenberg, 1991; Brunkhorst and Rollings, 1999; McGinnis et al, 1999, p204. See also Dryzek (1997) *The Politics of the Earth: Environmental Discourse.*

39  See Angelic Organics, 1547 Rockford Road, Caledonia, Illinois (www. angelicorganics.com).

40  ATTRA, 2000 (www.attra.org/attra-pub/csa.htm).

41  The attributes of box schemes are similar to North American CSAs, although CSAs generally expect a higher level of commitment from consumers. There has been no recent evaluation of box schemes in the UK; but Greg Pilley and colleagues of the Soil Association estimate that the 20 large schemes have up to 1200 customers each, and the 280 smaller ones an average of 200 customers, putting the total at 80,000 customers. Their judgement is that this may be optimistic, and thus 60,000 members is a more reasonable estimate. However, this appears to approach the number of members of all 1000 CSAs in the US, suggesting a need for a clear evaluation of the impacts of these schemes.

42  For more on the success of farmers' groups across all regions of the US, see documents from the Sustainable Agriculture and Rural Extension (SARE) Programme (1998) *Ten Years of SARE*, CSREES, USDA, Washington, DC (www.sare.org).

43  For US farmers' markets, see: www.ams.usda.gov/farmersmarket/facts.htm. See also Burns and Johnson (1999) *Farmers' Market Survey Report*, USDA (www.ams.usda.gov/directmarketing/wam024.htm), and Rominger, 2000.

44  For UK farmers' markets, see the National Association of Farmers' Markets (www.farmersmarkets.net).

45  For an excellent review of the wider policy and democratic issues surrounding food systems, see Lacy, 2000.

46  On food systems in North America, see Rod MacRae et al, 1993; 1999 pers comm; Wheeler et al, 1997; Mark Winne, pers comm, 1999.

47  Wheeler et al, 1997.

## Chapter 6   The Genetics Controversy

1  See Conway 2000; Royal Society et al, 2000.

2  Preliminary data in late 2001 from the International Service for the Acquisition of Agribiotech Applications (ISAAA) suggest that 50 million hectares of genetically modified crops were grown in 2001, up from 44.5 in the year 2000. In the year 2000, most genetically modified organisms were cultivated in the US (68 per cent), Argentina (23 per cent), Canada (7 per cent), with 25,000–100,000 hectares each in Australia, Mexico, Spain, and South Africa. There were also about 1000 hectares each in Bulgaria, France, Romania, Uruguay and Ukraine (Portugal grew a small amount in 1999, but then withdrew consent for 2000). In the UK, experimental field releases of genetically modified plants have occurred on 300 hectares. There are 400,000–500,000 hectares of genetically modified tobacco and cotton planted in China (see Chen, 2000; James, 2001; and www.isaaa.org). Of the total 44.5 million hectares planted worldwide in 2000, 58 per cent comprised soya; 23 per cent maize; 12 per cent cotton; and 6 per cent oilseed rape. The others include potato, squash and papaya.

3  See Stren and Alton, 1998.

4  For summaries of the contested views, see House of Lords, 1998; Royal Society, 1998; Nuffield Council on Bioethics, 1999; British Medical Association, 1999; Royal Society et al, 2000.

5  Bellagio Apomixis Declaration, 1998.

6  For a summary of environmental and health risks, see Rissler and Melon, 1996; Altieri, 1998; Pretty, 1998; House of Lords, 1999; Royal Society, 1998; Nuffield Council on Bioethics, 1999; BMA, 1999; ACRE 2000a, 2000b.

7  For a detailed discussion of the risks and benefits, see Pretty, 2001.

8  For more on gene flow, see Raybould and Gray, 1993; Chevré et al, 1997; DETR, 1999b.

9  See McPartlan and Dale, 1994; Gray and Raybould, 1998; BCPC, 1999; Young et al, 1999; ACRE, 2000b.

10  On potential gene flow in soils, see Gebhard and Smalla, 1998, 1999; ACRE, 2000b.

11  See Johnson, 2000.

12  For more on resistance, see Georghiou, 1986; Vorley and Keeney, 1998; Heap, 2000.

13  Royal Society, 1998.

14  For more on the indirect effects, see Birch et al, 1997; Hilbeck et al, 1998; Losey et al, 1999; Crecchio and Stotzky, 1998; Saxena et al, 1999.

15  For details of the research on monarch butterflies, see Losey et al, 1999; Monarch Butterfly Research Symposium, 1999; Jesse and Obrycki, 2000; Sears et al, 2001; Hemlich et al, 2001.

16  For more on biodiversity effects, see CRE, 1998; Royal Society, 1998; Johnson, 2000; Campbell et al, 1997; Pretty, 1998; Siriwardena et al, 1998; Mason, 1998.

17  For an overview of herbicide use, see Read and Bush, 1999; Dewar et al, 2000; Benbrook, 1999; Johnson, 2000.

18  See Royal Society, 1998; Nuffield Council on Bioethics, 1999.

19  *The Lancet*, 1999; for a summary, see Nuffield Council on Bioethics, 1999.

20  For more on the effects of antibiotic resistance, see House of Lords, 1998.

21  For more on the alternatives to antibiotic markers, see BMA, 1999; ACRE, 2000b.

22  In Royal Society, 1998.

23  See Grove-White et al, 1997; ESRC, 1999; US Senate Science Committee, 2000.

24  See Mary Shelley, 1818.

25  From O'Riordan, 1999.

26  US Senate Committee on Science, 2000.

27  Kloppenberg and Burrows, 1996; Altieri and Rosset, 1999a; Altieri, 1998.

28  Altieri and Rosset, 1999; Conway, 1997; Pretty 1995, 2000b; Gianessi and Carpenter, 1999; US Senate Science Committee, 2000.

29  For independent research on the effects of genetically modified organisms in the field, see Benbrook, 1999; ERS-USDA, 1999; Hyde et al, 2000; Minor et al, 1999; Oplinger et al, 1999; USDA, 1999; Conway, 2000; Elmore et al, 2001a, 2001b; Hal Willson, pers comm, 2000.

30  See UK House of Lords, 1999.

31  See Herdt, 1999; Hubbell and Welsh, 1999.

32  See Fowler and Mooney, 1990.

33  See Potrykus, 1999.

34  See McGloughlin, 1999; Altieri and Rosset, 1999a, 1999b.

35  For a summary of novel applications of genetic modification in developing countries, see Conway, 1997; Tanksley and McCouch, 1997; DFID, 1998; CGIAR, 2000; Royal Society et al, 2000; Winrock International, 2000; ISAAA, 2000.

36  See Pinto et al, 1998.

37  See Conway, 1997.

38  See Juma and Gupta, 1999.

39  See Pinstrup-Andersen, 1999.

# Chapter 7  Ecological Literacy

1  For more on the nature of traditional, see Posey, 1999.

2  See Lopez, 1998, p133.

3  See Scott (1998) *Seeing Like a State*, pp311, 332; Maturana and Varela (1992) *The Tree of Knowledge*. See also Capra, 1996.

4  See Kurukawa, 1992.

5  From Lopez, 1998, p144.

6  Ted Benton (1994) indicates that: *'There is now quite widespread agreement that . . . the dualistic opposites between subject and object, meaning and cause, mind and matter, harm and animal, and above all, culture (or society) and nature have to be rejected and transcended. The really difficult problems only start here, however.'*

7  Röling (2000) 'Gateway To the Global Garden'.

8  For a summary of social capital principles, see Pretty and Ward, 2001. For the main social capital literature of the past, see Tonnies, 1887; Kropotkin, 1902; Jacobs, 1961; Bourdieu, 1986; Coleman, 1988, 1990; Putnam, 1993, 1995.

9  For more on trust, see Gambetta, 1988; Fukuyama, 1995; Baland and Platteau, 1998.

10  For more on reciprocity, see Coleman, 1990; Putnam, 1993; Platteau, 1997.

11  For more on rights and responsibilities, see Taylor, 1982; Colins and Chippendale, 1991; Coleman, 1990; Elster, 1989; Etzioni, 1995.

12  For more on connectedness, see Uphoff, 1993; Cernea, 1993; Flora, 1998; Grootaert, 1998; Woolcock, 1998; Ward, 1998; Rowley, 1999; Barrett et al, 2001.

13  Firstly, there are local connections, comprising links between individuals within groups and communities. Local–local connections imply horizontal connections between groups within communities or between communities, which sometimes become new higher-level institutional structures. Then there are local–external connections, in which vertical connections between local groups and external agencies or organizations can be one way, usually top-down, or two way. External–external connections refer to horizontal connections between external agencies, leading to integrated approaches for collaborative partnerships. Finally, there are connections between individuals within external agencies.

14  See Pretty et al, 2001; Dobbs and Pretty, 2001a, 2001b.

15  See, for example, de los Reyes and Jopillo, 1986; Cernea, 1987, 1991; Uphoff, 1992; Pretty, 1995; Pretty et al, 1995; Bunch and López, 1996; Narayan and Pritchett, 1996; Röling and Wagemakers, 1997; Singh and Ballabh, 1997; Uphoff, 1998; Pretty and Hine, 2001.

16  Cernea, 1987.

17  On the 'dark side' of social capital, see Olson, 1965; Taylor, 1982; Knight, 1992; Portes and Landholt, 1996.

18  Barrett et al, 2001.

19  For more on the problem of free riders, see Grootaert, 1998; Dasgupta and Serageldin, 2000; Ostrom, 1998. For more on social innovation, see Boyte, 1995; Hamilton, 1995.

20  Long and Long, 1992; Röling and Jiggins, 1997; Pretty and Buck, 2002.

21  For a typology of seven different types of participation, see Pretty, 1995b.

22  Kang et al, 1984; Lal, 1989; Carter, 1995.

23  See Röling, 1997; Pretty, 1995b; Argyris and Schön, 1978; Habermas, 1987; Kenmore, 1999; Maturana and Varela, 1992.

24 See D Bromley, 1992.

25 On the hierarchy of the commons, see Johnson and Duchin, 2000; Buck, 1998. Two things are important about this hierarchy of commons. Firstly, actions at the lower levels influence the state and health of higher-level systems. Secondly, it is easier to take collective action at lower levels. The number of *stakeholders* with competing interests increases as we go up the hierarchy, which makes it more difficult to achieve collective action. But agreements at the higher levels can filter down to bring great changes. Ronald Oakerson has used a range of attributes to differentiate commons. The first is the degree of *jointness*, which refers to whether one person's use of the resource subtracts from its value for others. Such 'subtractability' may simply reduce the flow of benefits at one time, such as water or fish; or it may reduce the total yield of the common, perhaps changing it forever. The second is the degree of *exclusion*: how much access to the resource is controlled or restricted. If there is no exclusion, the resource is open access. If use is restricted to a defined group, then it is closed access. What is important is the system through which conditions for exclusion are applied. The third is the degree of *divisibility* of the commons: can the resource be divided among private property holders? Where should boundaries be drawn in order to define the resource and its users? The fourth is the *rules and decision-making arrangements* specified by a group of people. These include operational rules – how much should be taken or used, at what time and by whom, and the generalized norms by which individuals limit their actions in favour of the collective benefit. See Oakerson, 1992, p46

26 Singh and Bhattacharya, 1996.

27 This was predicted two decades ago by Olson, 1982.

28 For watershed groups, see Pretty, 1995b; IATP, 1998; Bunch, 2000; Hinchcliffe et al, 1999; F Shaxson, S Hocombe, A Mascaretti, pers comm, 1999; National Landcare Programme, 2000; Pretty and Frank, 2000.

29 For water users' groups, see de los Reyes and Jopillo, 1986; Bagadion and Korten, 1992; Ostrom, 1990; Uphoff, 1992; Cernea, 1993; Singh and Ballabh, 1997; Uphoff, 1998; Shah, 1998.

30 For microcredit groups, see Fernandez, 1992; Gibbons, 1996; Grameen Trust, *passim*.

31 For joint forest management, see Malla, 1997; Shrestha, 1997, 1998; SPWD, 1998; Raju, 1998; Poffenberger and McGean, 1998. Note than in India, the 25,000 joint forest management groups are managing 2.5 million hectares of forest, but the total amount of forest listed in gazetteers is 80 million hectares. There has been much progress, but still a long way to go.

32 Not every case of joint forest management (JFM) results in benefits for all local people, particularly if the forest department simply uses the name of JFM to exert control over local communities. Madhu Sarin recently documented the case of the village of Pakhi in Uttar Pradesh, where a women's group had sustainably managed a 240-hectare forest since the 1950s. But when the JFM programme was initiated in 1999, the local men formed the joint management group and ousted the women. Conflicts arose, and the forest department stepped

in to take back the key decisions. In the Uttarakhand region of Uttar Pradesh, there are 6000 community forests managed properly by communities, and half of households depend heavily upon these commons. The worst kind of development occurs when a good system is replaced by another (which turns out to be worse) in the name of sustainability. See Madhu Sarin (2001) *Disempowerment in the Name of Participatory Forestry? Village Forests Management in Uttarakhand.*

33  For integrated pest management farmer field-schools, see Kiss and Meerman, 1991; Matteson et al, 1992; Eveleens et al, 1996; van de Fliert, 1997; Kenmore, 1999; Desilles, 1999; Jones, 1999. See also Kenmore et al, 1984; Mangan and Mangan, 1998.

34  For farmers' groups, see Pretty, 1995a, 1995b; Harp et al, 1996; Oerlemans et al, 1997; van Weperen et al, 1997; van Veldhuizen et al, 1997; Just, 1998; Braun, 2000; Pretty and Hine, 2000. See Sue Heisswolf's thesis (2000) for the Rural Extension Centre, Gatton College, University of Queensland, for more on the value of social organization for agricultural change.

35  For more on the study of Iowan farmers, see Peter et al, 2000, p216. Monolegic implies a one-way connection, a transfer, instruction and the passing of information, whereas dialegic suggests two-way connection, an equal recognition of both partners, and connections between people–people and people–nature: see Bakhtin, 1981.

36  For more on CIALs, see Braun, 2000.

37  In her research in the southern state of Santa Caterina, Julia Guivant of the University of Florianópolis found substantial changes in women's welfare when families become involved in group production schemes. She says:

*Participation in production groups, whether involving agroindustry or not, allows the burden of agricultural production to be distributed between various families. This has led to important changes in the daily routine of the women, making it possible to share child-care in a way which would have been impossible with their husbands. Incorporating value added activities within these groups opens up new opportunities for women in the direction of greater empowerment: courses, direct contact with consumers, pride in their production, plans for future expansion.*

38  For more on the maturity of social capital in groups, see Bunch and López (1996) for Honduras and Guatemala; Bagadion and Korten (1991) for the Philippines; Uphoff (1998) for Sri Lanka; Krishna and Uphoff (1999) for Rajasthan, India; and Curtis et al (1999) for Australia.

39  See Pretty and Frank, 2000. The model identifies four distinct stages that relate the levels of total renewable assets to performance or outputs. These have been synthesized from a range of descriptive models that were developed for analysing changes in social capital manifested in groups and their life cycles (Mooney and Reiley, 1931; Handy, 1985; Pretty and Ward, 2001); for analysing types of participation between organizations and individuals (Pretty, 1995b; World Neighbors, 1999); for analysing changes in human capital manifested in phases of learning, knowing and world views through which individuals progress over time (Argyris and Schön, 1978; Habermas, 1987; Colins and Chippendale, 1991; Lawrence, 1999); for analysing changes in natural capital

during agricultural transformations (MacRae et al, 1993; Pretty, 1998); and for analysing adaptive management systems in terms of resilience, capital and connectivity (Holling, 1992).

40  See Ostrom, 1998.

41  See Röling and Wagemakers, 1998; Baland and Platteau, 1999; Dobbs and Pretty, 2001a, 2001b.

## Chapter 8    Crossing the Internal Frontiers

1  Intriguingly, the original of *The Man Who Planted Trees* was entitled by Giono as *The Man Who Planted Hope and Grew Happiness*.

2  See Jean Giono, 1954, pp34–37.

3  More than two decades before *The Man Who Planted Trees* was published, Jean Giono showed in *Second Harvest* how a Provençal village, again desertified, could be raised from the dead. In this story, a giant of a man, Panturle, has his hope rekindled by the arrival of Arsule, for whom he tills the soil and helps to remake the farm, cherry trees and meadows. Panturle more obviously suffers than the silent shepherd, and so when both he and the community are whole again, with children running and calling, and the fields full of crops, the sense of achievement is perhaps even greater:

> *Then, all of a sudden, standing there, he became aware of the great victory. Before his eyes passed the picture of the old earth, sullen and shaggy with its sour broom and knife-like grasses. . . He was standing in front of his fields. . . with his hands stretched down along his body, he stood motionless. He had won. It was over. He stood firmly placed in the earth like a pillar* (Giono, 1930, pp119–120).

4  A recent manifestation of the land ethic comes from E O Wilson's concept of biophilia. He defines biophilia as '*the connections that human beings subconsciously seek with the rest of life*', and argues that these are determined by a biological need. See Kellert and Wilson (1993) *The Biophilia Hypothesis*.

5  A Leopold (1932) 'Game and Wildlife Conservation', in *River and Other Essays*, quoted in Oelschlaeger, 1991, pp216–217.

6  See Howard (1940) *An Agricultural Testament*.

7  The wolf, of course, has added significance for humans; it was their symbiotic relationship with humans, guarding against other predators in return for scraps of food, that probably led to the domestication of the dog about 12,000–14,000 years ago. See Blench, 2001.

8  Leopold (1949) *A Sand County Almanac*, p129.

9  The Carpathian Ecoregional Initiative is one attempt to address rural development across the Carpathian region, and to draw in financial resources through ecotourism, thereby increasing the value of local natural assets. See Carpathian Large Carnivore Project and Carpathian Ecoregional Initiative.

10  See Wilson, 1988. See also Tilman, 2000; Myers et al, 2000; Wood et al, 2000; Bass et al, 2001; Stoll and O'Riordan, 2002. For a good overview, see

the whole issue of *Nature* (11 May 2000) on biodiversity with papers by Purvis and Hector; Gaston; McCann; Chapin et al; and Margules and Pressey, pp212–253.

11  See Orr, 2000. For more on design, see Orr, 2002.

12  See Ehrenfield, 2000, pp106–107, 100–111.

13  Abbotts Hall farm in Essex was purchased by the Essex Wildlife Trust with the support of the World Wide Fund for Nature, the Environment Agency, English Nature, The Wildlife Trusts and the Heritage Lottery Fund.

14  A measure of the difficulty of making these landscape changes to sea defences is given by the fact that the Essex Wildlife Trust had to obtain over 30 statutory consents, as well as approval through the formal planning process.

15  I am grateful to Eri Nakajima for pointing me to this story of ecological redesign, and for translating original government material from the Japanese.

16  McKean, 1985, pp67, 82

17  Devavaram et al, 1999; Rengasamy et al, 2000.

18  World Bank, 1995.

19  Schwarz and Schwarz, 1999; Smit et al, 1996; Rees, 1997.

20  See Garnett, 1996; National Society of Allotments and Leisure Gardeners (www.nsalg.co.uk).

21  For more on the psychological benefits of gardening, see Armstrong, 2000. See also Kaplan, 1973; McBey, 1985; WHO Regional Office for Europe, 2000.

22  Weissman, 1995a, 1995b.

23  Green Thumb tales from the field (www.cityfarmer.org/tales62.html).

# References

Abramovitz, J (1997) 'Valuing nature's services' in Brown L, Flavin C and French H (eds) *State of the World*. Worldwatch Institute, Washington, DC

ACC/SCN (2000) *4th Report on The World Nutrition Situation*. UN Administrative Committee on Coordination, Sub-Committee on Nutrition, in collaboration with IFPRI, United Nations, New York

ACRE (1998) *The Commercial Use of GM Crops in the UK: The Potential Wider Impact on Farmland Wildlife*. Advisory Committee on Releases to the Environment, Department of the Environment, Transport and the Regions (DETR), London

ACRE (2000a) *Annual Report 1999*. Advisory Committee on Releases to the Environment, Department of the Environment, Transport and the Regions (DETR), London

ACRE (2000b) *Gene Flow from Genetically Modified Crops*. Unpublished report. Advisory Committee on Releases to the Environment and the ACRE Secretariat. Department of the Environment, Transport and the Regions (DETR), London

Adams, J S and McShane, T O (1992) *The Myth of Wild Africa*. W W Norton and Co, New York

Agarwal, B (1997) 'Re-sounding the alert – gender, resources and community action', *World Development* 25(9), pp1373–1380

Agarwal B (1995) *Gender, Environment and Poverty Linkages in Rural India*. UNRISD Discussion Paper 62. UNRISD, Geneva

Altieri, M A (1995) *Agroecology: The Science of Sustainable Agriculture*. Westview Press, Boulder, Colorado

Altieri, M A (1998) *The Environmental Risks of Transgenic Crops: an agro-ecological assessment*. Department of Environmental Science, Policy and Management, University of California, Berkeley, California

Altieri, M A (1999) 'Enhancing the productivity of Latin American traditional peasant farming systems through an agro-ecological approach'. Paper for Conference on Sustainable Agriculture: New Paradigms and Old Practices? Bellagio Conference Centre, Italy, 26–30 April 1999

Altieri, M A and Rosset, P (1999a) 'Ten reasons why biotechnology will not ensure food security, protect the environment and reduce poverty in the developing world', *AgBioForum* 2(3–4), pp155–182 (www.agbioforum. org/Default/altieri.htm)

Altieri, M A and Rosset, P (1999b) 'Strengthening the case for biotechnology will not help the developing world: a response to McGloughlin', *AgBioForum* 2(3–4), pp226–236 (www.agbioforum.org/Default/altierireply.htm)

Angelic Organics, 1547 Rockford Road, Caledonia, Illinois (www.angelic organics.com)

Argyris, C and Schön, D (1978) *Organisational Learning.* Addison-Wesley, Reading, Massachusetts

Armstrong, D (2000) 'A survey of community gardens in upstate New York. Implications for health promotion and community development', *Health and Place* 6(4), pp319–327

Arnold, D (1996) *The Problem of Nature.* Blackwell Scientific, Oxford

Association for Better Husbandry (2000) Project reports and annual reports (www.ablh.org)

Astor, Viscount and Rowntree, B S (1945) *Mixed Farming and Muddled Thinking: An Analysis of Current Agricultural Policy.* Macdonald and Co, London

ATTRA (2000) Community-supported agriculture (www.attra.org/attra-pub/ csa.htm)

Australian National Drylands Salinity Programme (www.lwrrdc.gov.au/ndsp/ index.htm)

Avery, D (1995) *Saving the Planet with Pesticides and Plastic.* The Hudson Institute, Indianapolis

BAA (2000) *Annual Review and Handbook.* British Agrochemicals Association, Peterborough

Bagadion, B J and Korten, F F (1991) 'Developing irrigators' associations: a learning process approach' in Cernea, M M (ed) *Putting People First.* Oxford University Press, Oxford

Bailey, A P, Rehman, T, Park, J, Keatunge, J D H and Trainter, R B (1999) 'Towards a method for the economic evaluation of environmental indicators for UK integrated arable farming systems', *Agriculture, Ecosystems and Environment* 72, pp145–158

Bakhtin, M M (1981) *The Dialogic Imagination: Four Essays.* Edited by Holquist, M; translated by Emerson, C and Holquist, M. University of Texas Press, Austin, Texas

Baland, J-M and Platteau, J-P (1998) 'Division of the commons: a partial assessment of the new institutional economics of land rights', *American Journal of Agricultural Economics* 80(3), pp644–650

Balfour, E B (1943) *The Living Soil.* Faber and Faber, London

Barker, P (1998) 'Edge city' in Barnett, A and Scruton, R (eds) *Town and Country.* Jonathan Cape, London

Barnett, A and Scruton, R (eds) (1998) *Town and Country.* Jonathan Cape, London

Barrell, J (1980) *The Dark Side of the Landscape:* Cambridge University Press, Cambridge

Barrett, C and Grizzle, R (1999) 'A holistic approach to sustainability based on pluralism stewardship', *Environmental Ethics*, Spring, pp23–42

Barrett, C, Brandon, K, Gibson, C and Gjertsen, H (2001) 'Conserving tropical biodiversity amid weak institutions', *BioScience* 51(6), pp497–502

Bass, S, Hughes, C and Hawthorne, W (2001) 'Forests, biodiversity and livelihoods: linking policy and practice' in Koziell, I and Saunders, J (eds) *Living Off Biodiversity*. IIED, London

Baumol, W J and Oates, W E (1988) *The Theory of Environmental Policy*. Cambridge University Press, Cambridge

BCPC (1999) *Gene Flow and Agriculture: Relevance for Transgenic Crops*. Symposium Proceedings No 72. University of Keele and British Crop Protection Council, London

Beck, T (1994) *The Experience of Poverty: Fighting for Respect and Resources in Village India*. IT Publications, London

Beck, T and Ghosh, M (2000) 'Common property resources and the poor. Findings from West Bengal', *Economic and Political Weekly* 35(3), pp147–153

Beck, T and Naismith, C (2001) 'Building on poor people's capacities: the case of common property resources in India and West Africa', *World Development* 29(1), pp119–133

Beilin, R (2000) 'Sustaining a vision: recognising landscape futures'. Paper at Landcare International 2000 conference: Community participation in natural resource management. Melbourne, Australia

Bellagio Apomixis Declaration (1998) 27 April–1 May (www.billie.harvard.edu/apomixis)

Benbrook, C (1999) 'Evidence of the magnitude and consequences of the Roundup Ready soybean yield drag from university-based varietal trials in 1998', *Ag Bio Tech InfoNet Technical Paper* No 1. Sandpoint, Idaho (www.biotech-info.net/RR_yield_drag_98.pdf)

Bennett, D (1999) 'Stepping from the diagram: Australian Aboriginal cultural and spiritual values relating to biodiversity' in Posey, D (ed) *Cultural and Spiritual Values of Biodiversity*. IT Publications and UNEP, London

Benton, T (1994) 'Biology and social theory in the environmental debate' in Redclift, M and Benton, T (eds) *Social Theory and the Global Environment*. Routledge, London

Benton, T (1998) 'Sustainable development and the accumulation of capital: reconciling the irreconcilable?' in Dobson, A (ed) *Fairness and Futurity*. Oxford University Press, Oxford

Berkes, F (1998) 'Indigenous knowledge and resource management systems in the Canadian subarctic' in Berkes and Folke (eds) *Linking Social and Ecological Systems*. Cambridge University Press, Cambridge

Berkes, F and Folke, C (eds) (1998) *Linking Social and Ecological Systems*. Cambridge University Press, Cambridge

Berry, T (1998) *The Dream of the Earth*. Sierra Club Books, San Francisco

Berry, W (1977) *The Unsettling of America*. Sierra Club Books, San Francisco

Bignall, E M and McCracken, D I (1996) 'Low intensity farming systems in the conservation of the countryside', *Journal of Applied Ecology* 33, pp416–424

Birch, A, Geoghegan, I, Majerus, M, McNicol, J, Hackett, C, Gatehouse, A and Gatehouse, J (1997) 'Tri-trophic interactions involving pest aphids, predatory 2-spot ladybirds and transgenic potatoes expressing snowdrop lectin for aphid resistance', *Molecular Breeding* 5, pp75–83

Blench, R (2001) 'Why conserve livestock biodiversity?' in Koziell, I and Saunders, J (eds) *Living Off Biodiversity*. IIED, London

Bloch, M (1978) *French Rural History*. Routledge and Kegan Paul, London (1931 reprint)

Blum, J (1971) 'The European village as community: origins and functions', *Agricultural History Review* 45, pp157–178

Blythe, R (1999) *Talking about John Clare*. Trent Books, Nottingham

Bourdieu, P (1986) 'The forms of capital' in Richardson, J (ed) *Handbook of Theory and Research for the Sociology of Education*. Greenwood Press, Westport, Connecticut

Bowles, R and Webster, J (1995) 'Some problems associated with the analysis of the costs and benefits of pesticides', *Crop Protection* 14(7), pp593–600

Boyte, H (1995) 'Beyond deliberation: citizenship as public work'. Paper delivered at PEGS conference, 11–25 February 1995. Civic Practices Network (www.cpn.org)

Braun, A (2000) *The CIALs (Comité de Investigación Agrícultura Tropical) at a glance.* CIAT, Colombia

British Medical Association (1999) *The Impact of Genetic Modification on Agriculture, Food and Health*. BMA, London

Bromley, D W (ed) (1992) *Making the Commons Work*. Institute for Contemporary Studies Press, San Francisco, California

Brouwer, R (1999) *Market integration of agricultural externalities: a rapid assessment across EU countries*. Report for European Environment Agency, Copenhagen

Brummet, R (2000) 'Integrated aquaculture in Sub-Saharan Africa', *Environmental Development and Sustainability* 1(3–4), pp315–321

Bruner, A G, Gullison, R E, Rice, R E and de Fonseca, G A B (2001) 'Effectiveness of parks in protecting tropical biodiversity', *Science* 291, pp125–128

Brunkhorst, D J and Rollings, N M (1999) 'Linking ecological and social functions of landscapes: I. Influencing resource governance', *Natural Areas Journal* 19(1), pp57–64

Brunkhorst, D, Bridgewater, P and Parker, P (1997) 'The UNESCO biosphere reserve program comes of age: learning by doing; landscape models for sustainable conservation and resource use' in Hale, P and Lamb, D (eds) *Conservation Outside Nature Areas*. University of Queensland, Queensland, pp176–182

Buck, S J (1998) *The Global Commons*. Earthscan, London

Bunch, R (1999) 'Learning how to make the soil grow' in McDonald, M and Brown, K (eds) *Issues and Options in the Design of Soil and Water Conservation Projects*. Proceedings of a workshop held in Llandudno, Conwy, Wales, 1–3 February. University of Wales, Bangor

Bunch, R (2000) 'More productivity with fewer external inputs', *Environmental Development and Sustainability* 1(3–4), pp219–233

Bunch, R and López, G (1996) 'Soil recuperation in Central America: sustaining innovation after intervention', *Gatekeeper Series SA 55*, Sustainable Agriculture Programme, International Institute for Environment and Development, London

Burns, A and Johnson, D (1999) *Farmers' Market Survey Report*. USDA, Washington DC (www.ams.usda.gov/directmarketing/wam024.htm)

Butala, S (2000) 'Fields of broken dreams', *The Toronto Globe and Mail*, 4 March, Toronto

Butler-Flora, C and Flora, J L (1993) 'Entrepreneurial social infrastructure: a necessary ingredient', *The Annals of the American Academy of Political and Social Science* 529, pp48–55

Butler-Flora, C and Flora, J L (1996) 'Creating social capital' in Vitek, W and Jackson, W (eds) *Rooted in the Land: Essays on Community and Place*. Yale University Press, New Haven and London

Buzby, J C and Robert, T (1997) 'Economic costs and trade implications of microbial foodborne illness', *World Health Statistics Quarterly* 50(1–2), pp57–66

California Department of Food and Agriculture (1972–current) *Summary of Illnesses and Injuries Reported by Californian Physicians as Potentially Related to Pesticides*, 1972–current. Sacramento, California

Callicott, J B, Crowder, L B and Mumford, K (1999) 'Current normative concepts in conservation', *Conservation Biology* 13, pp22–35

Callicott, J B, Crowder, L B and Mumford, K (2000) 'Normative concepts in conservation biology. Reply to Willers and Hunter', *Conservation Biology* 14(2), pp575–578

Campbell, L H, Avery, M L, Donald, P, Evans, A D, Green, R E and Wilson, J D (1997) *A Review of the Indirect Effects of Pesticides on Birds*. Report No 227. Joint Nature Conservation Committee, Peterborough

Capra, F (1996) *The Web of Life*. HarperCollins, London

Carney, D (1998) *Sustainable Rural Livelihoods*. Department for International Development, London

Carpathian Ecoregional Initiative (2000) World Wide Fund for Nature, Austria, Vienna (www.carpathians.org)

Carpathian Large Carnivore Project (2000) *Annual Report*. Zarnesti, Romania (www.clcp.ro)

Carreck, N and Williams, I (1998) 'The economic value of bees in the UK', *Bee World* 79(3), pp115–123

Carson, R (1963) *Silent Spring*. Penguin Books, Harmondsworth

Carter, P (1987) *The Road to Botany Bay*. Faber and Faber, London

Carter, J (1995) *Alley Cropping: Have Resource Poor Farmers Benefited?* ODI Natural Resource Perspectives No 3, London

Cato, M P (1979) 'Di Agri Cultura' in Hooper, W D (revised Ash, H B) Marcus Porcius Cato *On Agriculture*, and Marcus Terentius Varro *On Agriculture*. Harvard University Press, Cambridge, Massachusetts

CDC (2001) 'Preliminary foodnet data on the incidence of foodborne illnesses', *MMWR Weekly* 50(13), pp241–246

Cernea, M M (1987) 'Farmer organisations and institution building for sustainable development', *Regional Development Dialogue* 8, pp1–24

Cernea, M M (1991) *Putting People First*. Oxford University Press, Oxford, second edition

Cernea, M M (1993) 'The sociologist's approach to sustainable development', *Finance and Development*, December, pp11–13

CGIAR (2000) *Promethean Science: Agricultural Biotechnology, the Environment and the Poor* (eds I Serageldin and G J Persley). CGIAR Secretariat. The World Bank, Washington, DC

Chen, Z L (2000) 'Transgenic food: need and safety'. Paper presented at OECD Edinburgh conference on The Scientific and Health Aspects of GM Foods, 28 February–1 March (www.oecd.org/subject/biotech/ed_prog_sum.htm)

Chevré, A-M, Eber, F, Baranger, A and Renard, M (1997) 'Gene flow from transgenic crops', *Nature* 389, p924

Clare, J (1993) *The Shepherd's Calendar* (ed Robinson, E, Summerfield, G and Powell, D). Oxford University Press, Oxford

Cleaver, K M and Schreiber, G A (1995) *The Population, Agriculture and Environment Nexus in Sub-Saharan Africa*. World Bank, Washington, DC

Cobb, D, Feber, R, Hopkins, A and Stockdale, L (1998) *Organic Farming Study*. Global Environmental Change Programme Briefing 17, University of Sussex, Falmer

Cobbett, W (1830) *Rural Rides* (ed Woodcock, G, 1967). Penguin Classics, Harmondsworth

Colchester, M (1997) 'Salvaging nature: indigenous peoples and protected areas' in Ghimire, K B and Pimbert, M (eds) *Social Change and Conservation*. Earthscan, London

Coleman, J (1988) 'Social capital and the creation of human capital', *American Journal of Sociology* 94, supplement S95-S120

Coleman, J (1990) *Foundations of Social Theory*. Harvard University Press, Harvard, Massachusetts

Colins, C J and Chippendale, P J (1991) *New Wisdom: The Nature of Social Reality*. Acorn Publications, Sunnybank, Queensland

Collier, W L, Wiradi, G and Soentoro (1973) 'Recent changes in rice harvesting methods: Some serious social implications', *Bulletin of Indonesian Economic Studies* 9(2), pp36–45

Common Ground (2000) *The Common Ground Book of Orchards*. Common Ground, London

Common, M (1995) *Sustainability and Policy*. Cambridge University Press, Cambridge

Conford, P (ed) (1988) *The Organic Tradition: An Anthology of Writing on Organic Farming*. Green Books, Bideford, Devon

Conway, G R (1997) *The Doubly Green Revolution*. Penguin, London

Conway, G R (2000) 'Crop biotechnology: benefits, risks and ownership'. Paper presented at OECD Edinburgh conference on The Scientific and Health Aspects of GM Foods, 28 February–1 March (www.oecd.org/subject/biotech/ed_prog_sum.htm)

Conway, G R and Pretty, J N (1991) *Unwelcome Harvest: Agriculture and Pollution*. Earthscan, London

Coop, P and Brunkhorst, D (1999) 'Triumph of the commons: age-old participatory practices provide lessons for institutional reform in the rural sector', *Australian Journal of Environmental Management* 6(2), pp48–56

Cooper, G (1996) 'Aldo Leopold and the values of nature' in Vitek, W and Jackson W (eds) *Rooted in the Land: Essays on Community and Place.* Yale University Press, New Haven and London

Cooper, N S (2000a) 'Speaking and listening to nature: ethics within ecology', *Biodiversity and Conservation* 9, pp1009–1027

Cooper, N S (2000b) 'How natural is a nature reserve? An ideological study of British nature conservation landscapes', *Biodiversity and Conservation* 9, pp1131–1152

Cosgrove, D and Daniels, S (1988) *The Iconography of Landscape.* Cambridge University Press, London

Costanza, R, d'Arge, R, de Groot, R, Farber, S, Grasso, M, Hannon, B, Limburg, K, Naeem, S, O'Neil, R V, Paruelo, J, Raskin, R G, Sutton, P and van den Belt, M (1997 and 1999). 'The value of the world's ecosystem services and natural capital', *Nature* 387, pp253–260; also in *Ecological Economics* 25(1), pp3–15

Countryside Agency (2001) *The State of the Countryside 2001.* Countryside Agency, Cheltenham

Crecchio, C and Stotzky, G (1998) 'Insecticidal activity and biodegradation of the toxin from *Bacillus thuringiensis subsp. kurstaki* bound to humic acids from soil', *Soil Biology and Biochemistry* 30, pp463–470

Cronon, W, Miles, G and Gitlin, J (eds) (1992) *Under an Open Sky. Rethinking America's Western Past.* W W Norton and Co, New York

Crosson, P (1995) 'Soil erosion estimates and costs', *Science* 269, pp461–464

Curtis, A, van Nouhays, M, Robinson, W and MacKay, J (1999) 'Exploring Landcare effectiveness using organisational theory'. Paper to International Symposium for Society and Natural Resources, University of Queensland, Brisbane, July 1999

Daily, G (ed) (1997) *Nature's Services: Societal Dependence on Natural Ecosystems.* Island Press, Washington, DC

Daily, G C, Matson, P A and Vitousek, P M (1997) 'Ecosystem services supplied by soil' in Daily, G C (ed) *Nature's Services.* Island Press, Washington, DC

Dasgupta, P and Serageldin, I (eds) *Social Capital: A Multiperspective Approach.* World Bank, Washington, DC

Davis, O and Kamien, M (1972) 'Externalities, information, and alternative collective action' in Dorfman, R and Dorfman, N (eds) *Economics of the Environment: Selected Readings.* W W Norton and Co, New York, pp69–87

Davison, M D, van Soest, J P, de Wit, G and De Boo, W (1996) *Financiële waardering van de milieuschade door de Nederlandse landbouw – een benadering op basis van de preventiekosten.* [Financial valuation of environmental hazard from Dutch agriculture – an approximation based on prevention costs]. Centre for Energy Conservation and Clean Technology (CE), Delft, The Netherlands

de Freitas, H (1999) 'Transforming microcatchments in Santa Caterina, Brazil' in Hinchcliffe, F, Thompson, J, Pretty, J, Guijt, I and Shah, P (eds) *Fertile Ground: The Impacts of Participatory Watershed Development.* IT Publications, London

de los Reyes, R and Jopillo, S G (1986) *An Evaluation of the Philippines Participatory Communal Irrigation Program.* Institute of Philippine Culture, Quezon City

Delgado, C, Rosegrant, M, Steinfield, H, Ehui, S and Courbois, C (1999) *Livestock to 2020: the next food revolution.* IFPRI Brief 61. International Food Policy Research Institute, Washington, DC

Desilles, S (1999) 'Sustaining and managing private natural resources: the way to step out of the cycle of high-input agriculture'. Paper for Conference on Sustainable Agriculture: New Paradigms and Old Practices? Bellagio Conference Centre, Italy, 26–30 April 1999

Department of the Environment, Food and Rural Affairs (DEFRA) (2000) *June Census data.* Economic and Statistics Group (www.defra.gov.uk/esg)

Department of the Environment, Transport and the Regions (DETR) (1997a) *Indicators of Sustainable Development for the UK.* Environmental Protection Statistics and Information Management Division (www.environment.detr.gov.uk/epsim/indcs/isd.htm)

DETR (1997b) *Digest of Environmental Statistics No 19* (www.detr.gov.uk/epsim/digest19/dig19-9.htm)

DETR (1998a) *The Environment in Your Pocket* (www.environment.detr.gov.uk/des20/pocket/env24.htm)

DETR (1998b) *Digest of Environmental Statistics No 20. UK Emissions of Greenhouse Gases* (www.environment.detr.gov.uk/des20/chapter1/)

DETR (1998c) *Good Practice Guide on Managing the Use of Common Land.* DETR, London

DETR (1999a) *Design of a Tax or Charge Scales for Pesticides.* DETR (now DEFRA), London

DETR (1999b) *Environmental risks of herbicide-tolerant oilseed rape. A review of the PGS hybrid oilseed rape.* Research Report No 15. DETR, London

Deutsch, S (1992) 'Landscape of enclaves' in Cronon, W, Miles, G and Gitlin, J (eds) (1992) *Under an Open Sky. Rethinking America's Western Past.* W W Norton and Co, New York

Devavaram, J, Arunothayam, E, Prasad, R and Pretty, J (1999) 'Watershed and community development in Tamil Nadu, India' in Hinchcliffe, F, Thompson, J, Pretty, J, Guijt, I and Shah, P (eds) *Fertile Ground: The Impacts of Participatory Watershed Development.* IT Publications, London

Dewar, A M, Haylock, L A, Bean, K M and May, A J (2000) 'Delayed control of weeds in glyphosate-tolerant sugar beet and the consequences on aphid infestation and yield', *Pest Management Science* 56(4), pp345–350

Department for International Development (DFID) Plant Science Research Programme (1998) *Rice Biotechnology.* University of Wales, Bangor

Diamond, J (1997) *Guns, Germs and Steel.* Vintage, London

Diop, A (2000) 'Sustainable agriculture: new paradigms and old practices? Increasing production with management of organic inputs in Senegal', *Environmental Development and Sustainability* 1(3–4), pp285–296

Dipera, K (1999) 'Botswana – Kaichela Dipera – Mukalahari' in Senanayake, R (ed) *Voices of the Earth,* pp121–167. In Posey, D (1999) *Cultural and Spiritual Values of Biodiversity.* IT Publications and UNEP, London

Djuretic, T, Wall, P G, Ryan, M J, Evans, H S, Adak, G K and Cowden, J M (1996) 'General outbreaks of infectious intestinal disease in England and Wales 1992–1994', *Communicable Disease Report* 6(4), R57–63

Dobbs, T L and Dumke, L M (1999) *Implications of 'Freedom to Farm' for Crop System Diversity in the Western Corn Belt and Northern Great Plains*. Economics Staff Paper 99-3. South Dakota State University, Brookings

Dobbs, T L and Pretty, J (2001a) *The United Kingdom's Experience with Agri-Environmental Stewardship Schemes: Lessons and Issues for the United States and Europe*. South Dakota State University Economics Staff Paper 2001-1 and University of Essex Centre for Environment and Society

Dobbs, T L and Pretty, J (2001b) *Future Directions for Joint Agricultural-Environmental Policies: Implications of the United Kingdom Experience for Europe and the United States*. South Dakota State University and University of Essex

Dobson, A (ed) (1999) *Fairness and Futurity*. Oxford University Press, Oxford

Domestic Animal Diversity Information System (DADIS): www.dad.fao.org/cgi-dad/$cgi_dad.exe/summaries

Donaldson, J (1854) *Agricultural Biography. Life and Writings of the British Authors of Agriculture*. London

Drinkwater, L E, Wagoner, P and Sarrantonio, M (1998) 'Legume-based cropping systems have reduced carbon and nitrogen losses', *Nature* 396, pp262–265

Dryzek, J (1997) *The Politics of the Earth: Environmental Discourse*. Oxford University Press, New York and Oxford

Dubois, D, Fried, P M, Deracuasaz, B and Lehman, H (2000) 'Evolution and instruments for the implementation of a program for whole farm environmental management in Switzerland'. OECD Workshop on Adoption of Technologies for Sustainable Farming Systems, The Netherlands, July 2000. OECD, Paris COM/AGR/CA/ENV/EPOC (2000) 65

Duffy, R (2000) *Killing for Conservation: Wildlife Policy in Zimbabwe*. James Currey, Oxford

Dumke, L M and Dobbs, T L (1999) *Historical Evolution of Crop Systems in Eastern South Dakota: Economic Influences*. Economics Research Report 99-2. South Dakota State University, Brookings

DuPuis, E M and Vandergeest, P (eds) *Creating Countryside: The Politics of Rural and Environmental Discourse*. Temple University Press, Philadelphia

Durrenberger, E P and Thu, K M (1996) 'The expansion of large scale hog farming in Iowa: the applicability of Goldschmidt's findings fifty years later', *Human Organisation* 55(4), pp409–415

*Ecological Economics* (1999), 25(1). Special issue devoted to Costanza et al (1997) paper, with 12 responses (Ayres; Daly; El Serafy; Herendeen; Hueting et al; Norgaard and Bode; Opschoor; Pimentel; Rees; Templet; Toman; and Turner et al), and a reply from Costanza et al

European Environment Agency (EEA) (1996) *Environmental Taxes: Implementation and Environmental Effectiveness*. Environmental Issues Series No 1. EEA, Copenhagen

EEA (1998) *Europe's Environment: The Second Assessment. Report and Statistical Compendium.* EEA, Copenhagen

EEA (1999) *Annual European Community Greenhouse Gas Inventory 1990-1996.* Technical Report No 19, EEA, Copenhagen

Ehrenfield, D (2000) 'War and peace and conservation biology', *Conservation Biology* 14(1), pp105–112

Eisinger, P K (1998) *Towards an End of Hunger in America.* Brookings Institution Press, Washington, DC

Ekins, P (1999) 'European environmental taxes and charges: recent experience, issues and trends', *Ecological Economics* 31, pp39–62

Elmore, R W, Roeth, F W, Klein, R N, Knezevic, S V, Martin, A, Nelson, L A and Shapiro, C A (2001b) 'Glyphosate-resistant soybean cultivar response to glyphosate', *Agronomy Journal* 93, pp404–407

Elmore, R W, Roeth, F W, Nelson, L A, Shapiro, C A, Klein, R N, Knezevic, S V and Martin, A (2001a) 'Glyphosate-resistant soybean cultivar yields relative to sister lines', *Agronomy Journal* 93, pp408–412

Elster, J (1989) *The Cement of Society: A Study of Social Order.* Cambridge University Press, Cambridge

Engels, F (1956) *The Peasant War in Germany.* Progress Publishers, Moscow

English Nature (1999) *Common Land: unravelling the mysteries.* English Nature, Peterborough

English Tourist Council (ETC) (2000) *United Kingdom Tourist Statistics 1999.* ETC, London

Environment Agency (EA) (1998) *Aquatic Eutrophication in England and Wales: a proposed management strategy.* EA, Bristol

Environment Agency of Japan (1999) *One Summer in Satochi.* Tokyo, Japan

Ernle, Lord (Prothero, R) (1912) *English Farming. Past and Present.* Longmans, Green and Co, London

Escalada, M M, Heong, K L, Huan, N H and Mai, V (1999) 'Communication and behaviour change in rice pest management: the case of using mass media in Vietnam', *Journal of Applied Communications* 83(1), pp7–26

ESRC (1999) *The Politics of GM Food.* Special Briefing No 5, October 1999. ESRC Global Environmental Change Programme, University of Sussex, Brighton

Etzioni, A (1995) *The Spirit of Community: Rights, Responsibilities and the Communitarian Agenda.* Fontana Press, London

Evans, G E (1960) *The Horse and the Furrow.* Faber, London

Evans, H S, Madden, P, Douglas, C, Adak, G K, O'Brien, S J, Djuretic, T, Wall, P G and Stanwell-Smith, R (1998) 'General outbreaks of infectious disease in England and Wales 1995–1996', *Communicable Disease and Public Health* 1(3), pp165–171

Evans, R (1995) *Soil Erosion and Land Use: Towards a Sustainable Policy.* Cambridge Environmental Initiative, University of Cambridge, Cambridge

Eveleens, K G, Chisholm, R, van de Fliert, E, Kato, M, Thi Nhat, P and Schmidt, P (1996) *Mid Term Review of Phase III Report – The FAO Intercountry Programme for the Development and Application of Integrated Pest Management Control in Rice in South and South East Asia.* GCP/RAS/145-147/NET-AUL-SWI. FAO, Rome

Evers, L and Molina, F S (1987) *Maso/Bwikam/Yaqui Deer Songs: A Native American Poetry*. Sun Track and University of Arizona Press, Tucson, Arizona

Ewel, K C (1997) 'Water quality improvement in wetlands' in Daily, G (ed) (1997) *Nature's Services: Societal Dependence on Natural Ecosystems*. Island Press, Washington, DC

Food and Agriculture Organization (FAO) (1999) *Cultivating Our Futures: Taking Stock of the Multifunctional Character of Agriculture and Land*. FAO, Rome

FAO (1999) *The Future of Our Land*. FAO, Rome

FAO (2000a) *Agriculture: Towards 2015/30*. Global Perspective Studies Unit, FAO, Rome

FAO (2000b) *Food Security Programme* (www.fao.org/sd/FSdirect/FSPintro. htm)

FAO (2001) *FAOSTAT Database*. FAO, Rome

FAO (2001) *Animal agriculture in the EU. Some elements for a way forward*. Animal Production and Health Division, FAO, Rome

FAO/UNEP (2000) 'World watch list for domestic animal diversity' (www.fao. org/dad-is)

Fernandez, A (1992) *The MYRADA Experience: Alternate Management Systems for Savings and Credit of the Rural Poor*. MYRADA, Bangalore

FiBL (2000) *Organic Farming Enhances Soil Fertility and Biodiversity. Results from a 21 year field trial*. FiBL Dossier 1 (August). Research Institute of Organic Agriculture (FiBL), Zurich

Fleischer, G and Waibel, H (1998) 'Externalities by pesticide use in Germany'. Paper presented to Expert Meeting: The Externalities of Agriculture: What do we Know? EEA, Copenhagen, May 1998

Flora, J L (1998) 'Social capital and communities of place', *Rural Sociology* 63(4), pp481–506

Foreman, R T (1997) *Land Mosaics: The Ecology of Landscapes and Regions*. Cambridge University Press, Cambridge

Foster, V, Bateman, I J and Harley, D (1997) 'Real and hypothetical willingness to pay for environmental preservation: a non-experimental comparison', *Journal of Agricultural Economics* 48(1), pp123–138

Fowler, C and Mooney, P (1990) *The Threatened Gene: Food, Policies and the Loss of Genetic Diversity*. The Lutterworth Press, Cambridge

Fuglie, K, Narrod, C and Neumeyer, C (2000) 'Public and private investment in animal research' in Fuglie, K and Schimmelpfenning, D (eds) *Public-Private Collaboration in Agricultural Research*. Iowa State University Press, Des Moines, Iowa

Fukuoka, M (1985) *The Natural Way of Farming* (translated by F R Metreaud). Japan Publications Inc, Tokyo and New York

Fukuyama, F (1995) *Trust: The Social Values and the Creation of Prosperity*. Free Press, New York

Funes, F (2001) 'Cuba and sustainable agriculture'. Paper presented to St James's Palace conference: Reducing Poverty with Sustainable Agriculture, 15 January

Funes, F, García, L, Bourque, M, Pérez, N and Rosset, P (2002) *Sustainable Agriculture and Resistance*. Food First Books, Oakland, California

Gadgil, M and Guha, R (1992) *This Fissured Land: An Ecological History of India*. Oxford University Press, New Delhi

Gadgil, M (1998) 'Grassroots conservation practices: revitalizing the traditions' in Kothari, A, Pathak, N, Anuradha, R V and Taneja, B (eds) *Communities and Conservation: Natural Resource Management in South and Central Asia*. Sage Publications, New Delhi

Gambetta, D (ed) (1988) *Trust: Making and Breaking Cooperative Relations*. Blackwell Scientific, Oxford

Gapor, S A (2001) *Rural Sustainability in Sarawak. The role of adat and indigenous knowledge in promoting sustainable sago production in the coastal areas of Sarawak*. PhD thesis, University of Hull, Hull

Garkovich, L, Bokemeier, J and Foote, B (1995) *Harvest of Hope*. University of Kentucky Press, Lexington

Garnett, T (1996) *Growing Food in Cities: A report to highlight and promote the benefits of urban agriculture in the UK*. SAFE Alliance and National Food Alliance, London

Garreau, J (1992) *Edge City – Life on the New Frontier*. Anchor Books, New York

Gebhard, F and Smalla, K (1998) 'Transformation of *Acinetobacter* sp. strain BD413 by transgenic sugar beet DNA', *Applied and Environmental Microbiology* 64(4), pp1550–1554

Gebhard, F and Smalla, K (1999) 'Monitoring field releases of genetically modified sugar beets for persistence of transgenic plant DNA and horizontal gene transfer', *FEMS Microbiology Ecology* 28, pp261–272

Geertz, C (1993) *Local Knowledge. Further Essays in Interpretive Anthropology*. Fontana Press, London

Georghiou, G P (1986) 'The magnitude of the problem' in National Research Council, *Pesticide Resistance, Strategies and Tactics*. National Academy Press, Washington, DC

Ghimire, K and Pimbert, M (1997) *Social Change and Conservation*. Earthscan, London

Gianessi, L P and Carpenter, J E (1999) *Agricultural Biotechnology: Insect Control Benefits*. National Center for Food and Agricultural Policy, Washington, DC

Gibbons, D S (1996) 'Resource mobilisation for maximising MFI outreach and financial self-sufficiency'. Issues Paper No 3 for Bank-Poor 1996, 10–12 December, Kuala Lumpur

Grameen Bank at www.grameen.com or www.grameen-info.org

Giono, J (1930) *Second Harvest*. Harvill Press, London (1999 edition)

Giono, J (1954) *The Man Who Planted Trees*. Peter Owen, London (1985 edition)

Goin, P (1992) *Humanature*. University of Texas Press, Harrisonberg, Virginia

Goldschmidt, W (1978) (first published in 1946). *As You Sow: Three Studies in the Social Consequences of Agri-Business*. Allanheld, Monclair, New Jersey

Gómez-Pompa, A and Kaus, A (1992) 'Taming the wilderness myth', *Bioscience* 42(4), pp271–279

Grainger, M (ed) (1993) *The Natural History Prose Writings of John Clare*. Clarendon Press, Oxford

Gray, A (1999) 'Indigenous peoples, their environments and territories' in Posey D (ed) (1999) *Cultural and Spiritual Values of Biodiversity*. IT Publications and UNEP, London

Gray, A and Raybould, A (1998) 'Reducing transgene escape routes', *Nature* 312, pp653–654

Greene, C and Dobbs, T (2001) *Organic wheat production in the US*. Economic Research Service (USDA), wheat yearbook. USDA, Washington, DC, pp31–37

GreenThumb, 'Tales from the field' (www.cityfarmer.org/tales62.html)

Gren, I M (1995) 'Costs and benefits of restoring wetlands: two Swedish case studies', *Ecological Engineering* 4, pp153–162

Grimes, B (1996) *Ethnologue: Languages of the World*. 12th Edition. Summer Institute of Linguistics, Dallas, Texas

Grootaert, C (1998) 'Social capital: the missing link'. World Bank Social Capital Initiative Working Paper No 5, Washington, DC

Grove, R H (1990) 'Colonial conservation, ecological hegemony and popular resistance: towards a global synthesis' in MacKenzie, J M (ed) *Imperialism and the Natural World*. Manchester University Press, Manchester.

Grove-White, R, Macnaughton, P, Mayer, S and Wynne, B (1997) *Uncertain Worlds: GMOs, Food and Public Attitudes in Britain*. CSEC, Lancaster University, Lancaster

Habermas, J (1987) *Theory of Communicative Action: Critique of Functionalist Reason*. Volume II. Polity Press, Oxford

Hamilton, N A (1995) *Learning to Learn with Farmers*. PhD thesis, Wageningen Agricultural University, The Netherlands

Hamilton, P (1998) *Goodbye to Hunger: A study of farmers' perceptions of conservation farming*. ABLH, Nairobi, Kenya

Handy, C (1985) *Understanding Organisations*. Penguin Books, Harmondsworth

Hanley, N and Oglethorpe, D (1999) 'Toward policies on externalities from agriculture: an analysis for the European Union'. Paper presented at American Agricultural Economics Association Annual Meeting, Nashville, Tennessee

Hanley, N, MacMillan, D, Wright, R E, Bullock, C, Simpson, I, Parrison, D and Crabtree, R (1998) 'Contingent valuation versus choice experiments: estimating the benefits of environmentally sensitive areas in Scotland', *Journal of Agricultural Economics* 49(1) pp1–15

Hardin, G (1968) 'The tragedy of the commons', *Science* 162, pp1243–1248

Harp, A, Boddy, P, Shequist, K, Huber, G and Exner, D (1996) 'Iowa, USA: An effective partnership between the Practical Farmers of Iowa and Iowa State University' in Thrupp, L A (ed) *New Partnerships for Sustainable Agriculture*. WRI, Washington, DC

Harrington, G (2000) *The Future of the Pig Industry – Local and Global*. Silcock Fellowship for Livestock Research Report 7, Harper Adams University College, Newport

Harrison, P F and Lederberg, J (eds) (1998) *Antimicrobial Resistance: issues and options*. National Academy Press, Washington, DC

Hartridge, O and Pearce, D (2001) *Is UK Agriculture Sustainable? Environmentally Adjusted Economic Accounts.* CSERGE, University College, London

Hasselstrom, L (1997) 'Addicted to Work' in Vitek, W and Jackson, W (eds) *Rooted in the Land: Essays on Community and Place.* Yale University Press, New Haven and London

Havelaar, A M, de Wit, M A S, van Kuningeveld, R and van Kempen, E (2000) 'Health burden in the Netherlands due to infection with thermophilic Campylobacter species', *Epidemiological Infection* 125, pp505–522

Heap, I (2000) 'International survey of herbicide resistant weeds' at: www.weedscience.org/summary/countrysum.asp

Heffernan, W, Henrickson, M and Gronkski, R (1999) *Consolidation in the Food and Agriculture System.* Report to the National Farmers Union. University of Missouri, Columbia (www.nfu.org)

Heimlich, R E, Wiebe, K D, Claasen, R, Gadsby, D and House, R M (1998) *Wetlands and Agriculture: Private Interests and Public Benefits.* Resource Economics Division. Economic Research Service, USDA. Agricultural Economics Report No 765. USDA, Washington, DC

Heisswolf, S (2001) *Building social capital in horticulture: Factors impacting on the establishment and sustainability of farmer groups.* QM742 Extension Dissertation. Rural Extension Centre, Gatton College, University of Queensland, Queensland

Henao, J and Baanante, C (1999) *Nutrient depletion in the agricultural soils of Africa.* IFPRI Brief 62, IFPRI, Washington, DC

Heong, K L, Escalada, M M, Huan, N H and Mai, V (1999) 'Use of communication media in changing rice farmers' pest management in the Mekong Delta, Vietnam', *Crop Management* 17(5), pp413–425

Herdt, R (1999) 'Enclosing the global plant genetic commons'. Paper presented at Institute for International Studies, Stanford University. The Rockefeller Foundation, New York

Hilbeck, A, Baumgartner, M, Fried, P M and Bigler, F (1998) 'Effects of transgenic *Bt* corn-fed prey on mortality and development time of immature *Chrysoperla carnea Neuroptera: Chrysopidae*', *Environmental Entomology* 27, pp460–487

Hinchcliffe, F, Thompson, J, Pretty, J, Guijt, I and Shah, P (eds) (1999) *Fertile Ground: The Impacts of Participatory Watershed Development.* IT Publications, London

Hoddinott, J (1999) *Operationalising Household Food Security in Development Projects.* IFPRI, Washington, DC

Holling, C S (1992) 'Cross-scale morphology, geometry and dynamics of ecosystems', *Ecological Monographs* 2(4), pp447–502

Holling, C S and Meffe, M (1996) 'Commons and control and the pathology of natural resource management', *Conservation Biology* 10(2), pp328–337

Hoskins, W G (1955) *The Making of the English Landscape.* Penguin Books, London

House of Lords Select Committee on Science and Technology (1998) *Seventh Report: Resistance to antibiotics and other microbiological agents.* HMSO, London

House of Lords Select Committee on the European Communities (1999) *EC Regulation of Genetic Modification in Agriculture.* HMSO, London

Howard, A (1940) *An Agricultural Testament.* Faber, London

Health and Safety Executive (HSE) (1998a) *Pesticides Incidents Report 1997/8.* HSE, Sudbury

HSE (1998b) *Pesticide Users and their Health: Results of HSE's 1996/7 Feasibility Study* (www.open.gov.uk/hse/hsehome.htm)

Hubbell, B J and Welsh, R (1998) 'Transgenic crops: engineering a more sustainable future?' *Agriculture and Human Values* 15, pp43–56

Humphries, J (1990) 'Enclosures, common rights and women: the proliterianization of families in the late eighteenth and early nineteenth centuries', *Journal of Economic History* 50(1), pp17–42

Hunter, M L Jr (2000) 'Refining normative concepts in conservation', *Conservation Biology* 14(2), pp573–574

Hutcheon, L (1989) *The Politics of Postmodernism.* Routledge, London

Huxley, E (1960) *A New Earth: An Experiment in Colonialism.* Chatto and Windus, London

Hyde, J, Martin, M A, Preckel, P V and Edwards, C R (2000) 'The economics of *B.t.* corn: adoption and implications'. Department of Agricultural Economics, Purdue University (www.agcom.purdue.edu/AgCom/Pubs/ID-219/ID-219.html)

IATP (1998) 'Farmer-managed watershed program' (www.iatp.org)

ISAAA (2001) 'ISAAA in Brief' (www.isaaa.org)

Iyengar, S and Shukla, N (1999) *Common property land resources in India: some issues in regeneration and management.* Gujarat Institute of Development Research, Ahmedabad

Jacobs, J (1961) *The Life and Death of Great American Cities.* Random House, London

James, C (2001) *Global status of commercialised transgenic crops, 2000.* ISAAA Briefs No 21-2001 (www.isaaa.org/briefs/Brief 21.htm)

Jarass, L and Obermair, G M (1997) *More Jobs, Less Tax Evasion, Cleaner Environment.* Universität Regensburg, Germany (www.suk.fh-wiesbaden.de/personen/jarass/manuskript5.html)

Jennings, F (1984) *The Ambiguous Iroquois Empire.* W W Norton and Co, New York

Jesse, L C H and Obrycki, J J (2000) 'Field deposition of *Bt* transgenic corn pollen: lethal effects on the monarch butterfly', *Oceologia* 125, pp241–248

Jodha, N S (1988) 'Poverty debate in India: a minority view', *Economic and Political Weekly*, November, pp2421–2428

Jodha, N S (1990) 'Common property resources and rural poor in dry regions of India', *Economic and Political Weekly* 21, pp1169–1181

Jodha, N S (1991) *Rural common property resources: a growing problem.* Sustainable Agriculture Programme Gatekeeper Series No 24, IIED, London

Johnson, B and Duchin, F (2000) 'The case for the global commons' in Harris J (ed) *Rethinking Sustainability: Power, Knowledge and Institutions.* University of Michigan Press, Ann Arbor

Johnson, B (2000) *Problems of plant conservation in agricultural landscapes: can biotechnology help or hinder?* Unpublished report. English Nature, Peterborough

Jones, K (1999) 'Integrated pest and crop management in Sri Lanka'. Paper for Conference on Sustainable Agriculture: New Paradigms and Old Practices? Bellagio Conference Centre, Italy, 26–30 April 1999

Juma, C and Gupta, A (1999) 'Safe use of biotechnology', *IFPRI 2020 Vision Focus 2*, Brief 6 of 10. International Food Policy Research Institute, Washington, DC

Just, F (1998) 'Do soft regulations matter?' Paper presented to Expert Meeting: The Externalities of Agriculture: What do we Know? European Environment Agency, Copenhagen, May 1998

Kaeferstein, F K, Motarjeni, Y and Bettcher, D W (1997) 'Foodborne disease control', *Emerging Infectious Diseases* 3, pp503–516

Kang, B T, Wilson, G F and Lawson, T L (1984) *Alley Cropping: A Stable Alternative to Shifting Agriculture*. IITA, Ibadan

Kaplan, D (1973) 'Some psychological benefits of gardening', *Environment and Behaviour* 5, pp145–161

Kato, Y, Yokohari, M and Brown, R D (1997) 'Integration and visualisation of the ecological value of rural landscapes in maintaining the physical environment of Japan', *Landscape and Urban Planning* 39, pp69–82

Katz, E G (2000) 'Social capital and natural capital: a comparative analysis of land tenure and natural resource management in Guatemala', *Land Economics* 76(1), pp114–132

Keeny, D and Muller, M (2000) *Nitrogen and the Upper Mississippi River*. Institute of Agriculture and Trade Policy, Minneapolis

Kellert, S and Wilson, E O (1993) *The Biophilia Hypothesis*. Island Press/Shearwater, Washington, DC

Kenmore, P E, Carino, F O, Perez, C A, Dyck, V A and Gutierrez, A P (1984) 'Population regulation of the brown planthopper within rice fields in the Philippines', *Journal of Plant Protection in the Tropics* 1(1), pp19–37

Kenmore, P E (1999) 'Rice IPM and farmer field schools in Asia'. Paper for Conference on Sustainable Agriculture: New Paradigms and Old Practices? Bellagio Conference Centre, Italy, 26–30 April 1999

Khan, Z R, Pickett, J A, van den Berg, J and Woodcock, C M (2000) 'Exploiting chemical ecology and species diversity: stem borer and *Striga* control for maize in Africa', *Pest Management Science* 56(1), pp1–6

Khare, A (1998) 'Community-based conservation in India' in Kothari, A, Pathak, N, Anuradha, R V and Taneja, B (1998) *Communities and Conservation: Natural Resource Management in South and Central Asia*. Sage, New Delhi

King, F H (1911) *Farmers of Forty Centuries: Permanent Agriculture in China, Korea and Japan*. Rodale Press, Pennsylvania

Kiss, A and Meerman, F (1991) *Integrated Pest Management in African Agriculture*. World Bank Technical Paper 142, World Bank, Washington, DC

Klijn, J and Vos, W (eds) (2000) *From Landscape Ecology to Landscape Science*. Kluwer Academic Publishers, Dordrecht

Kline, D (1996) 'An Amish Perspective' in Vitek, W and Jackson, W (eds) *Rooted in the Land: Essays on Community and Place*. Yale University Press, New Haven and London

Kloppenberg, J (1991) 'Social theory and the de/reconstruction of agricultural science: a new agenda for rural sociology', *Sociologia Ruralis* 32(1), pp519–548

Kloppenberg, J and Burrows, B (1996) 'Biotechnology to the rescue: twelve reasons why biotechnology is incompatible with sustainable agriculture', *The Ecologist* 26(2), pp61–67

Knight, J (1992) *Institutions and Social Conflict.* Cambridge University Press, Cambridge

Knutson, R, Penn, J and Flinchbaugh, B (1998) *Agricultural and Food Policy*, Fourth Edition. Prentice Hall, Upper Saddle River, New Jersey

Koohafkan, P and Stewart, B A (2001) *Water Conservation and Water Harvesting in Cereal-Producing Regions of the Drylands.* FAO, Rome

Kothari, A, Pande, P, Singh, S and Dilnavaz, R (1989) *Management of National Parks and Sanctuaries in India.* Indian Institute of Public Administration, New Delhi

Kothari, A, Pathak, N, Anuradha, R V and Taneja, B (1998) *Communities and Conservation: Natural Resource Management in South and Central Asia.* Sage, New Delhi

Kovaleski, S F (1999) 'Cuba goes green: government-run vegetable gardens sprout in cities across island' at: www.cityfarmer.org/CubaGreen.html

Krebs, J R, Wilson, J D, Bradbury, R B and Siriwardena, G M (1999) 'The second silent spring?' *Nature* 400, pp611–612

Krishna, A and Uphoff, N (1999) *Operationalising social capital: explaining and measuring mutually beneficial collective action in Rajasthan, India.* Cornell University, Cornell

Kropotkin, P (1902) *Mutual Aid.* Extending Horizon Books, Boston (1955 edition)

Kurokawa, K (1991) *Intercultural Architecture. The Philosophy of Symbiosis.* Academy Editions, London

Lacy, W B (2000) 'Empowering communities through public work, science and local food systems: revisiting democracy and globalisation', *Rural Sociology* 65(1), pp3–26

Lal, R (1989) 'Agroforestry systems and soil surface management of a Tropical Alfisol. I: Soil moisture and crop yields', *Agroforestry Systems* 8, pp7–29

Lampkin, N (1996) *Impact of EC Regulation 2078/92 on the development of organic farming in the European Union.* Welsh Institute of Rural Affairs, University of Wales, Aberystwyth, Dyfed

Lampkin, N and Midmore, P (2000) 'Changing fortunes for organic farming in Europe: policies and prospects'. Paper presented for Agricultural Economics Society Annual Conference, Manchester, UK

Lampkin, N and Padel, S (eds) (1994) *The Economics of Organic Farming: An International Perspective.* CAB International, Wallingford

Landers, J (1999) 'Policy and organisational dimensions of the process of transition toward sustainable intensification in Brazilian agriculture'. Paper presented for Rural Week, The World Bank, 24–26 March, Washington, DC

Landers, J N, De C Barros, G S-A, Manfrinato, W A, Rocha, M T and Weiss, J S (2001) 'Environmental benefits of zero-tillage in Brazil – a first approximation' in Garcia Torres, L, Benites, J and Martinez Vilela, A (eds) *Conservation Agriculture – A Worldwide Challenge.* Volume I. XUL, Cordoba, Spain

Lane, C (1990) *Barabaig Natural Resource Management: sustainable land use under threat of destruction.* UNRISD Discussion Paper 12, Geneva

Lane, C (1993) 'The state strikes back: extinguishing customary rights to land in Tanzania' in *Never Drink from the Same Cup*. Proceedings of the Conference on Indigenous Peoples in Africa. CDR/IWIGIA Document 72, Denmark

Lane, C and Pretty, J (1990) *Displaced Pastoralists and Transferred Wheat Technology in Tanzania*. Sustainable Agriculture Programme Gatekeeper Series SA20. IIED, London

Lang, T, Heasman, M and Pitt, J (1999) *Food, Globalisation and a New Public Health Agenda*. International Forum on Globalization, San Francisco

Lang, T, Barling, D and Caraher, M (2001) 'Food, social policy and the environment: towards a new model', *Social Policy and Administration* 35(5), pp538–558

Lawrence, D (1999) 'Stages of adult learning'. Mimeo. Queensland Department of Primary Industries, Australia

Leach, M and Mearns, R (1996) *The Lie of the Land*. Routledge, London

Lease, G (1995) 'Introduction: Nature under fire' in Soulé, M E and Lease, G (eds) *Reinventing Nature? Response to Postmodern Deconstruction*. Island Press, Washington, DC

Leopold, A (1932) 'Game and Wildlife Conservation' in *River and Other Essays*, quoted in Oelschlaeger, M (1991) *The Idea of Wilderness*. Yale University Press, New Haven, pp216–217

Leopold, A (1949) *A Sand County Almanac and Sketches Here and There*. Oxford University Press, London and New York (1974 edition)

Lewis, G M (1988) 'Rhetoric of the western interior: modes of environmental description in American promotional literature of the 19th century' in Cosgrove, D and Daniels, S (1988) *The Iconography of Landscape*. Cambridge University Press, London

Lewis, O (1964) *Pedro Martinez*. Alfred Knopf, New York

Lindqvist, R, Andersson, Y, Lindback, J, Wegscheider, M, Eriksson, Y, Tidestrom, L, Lagerqvist-Widh, A, Hedlund, K-O, Lofdahl, S, Svensson, L and Norinder, A (2001) 'A one-year study of foodborne illnesses in the municipality of Uppsala, Sweden', *Emerging Infectious Diseases* (Centre for Disease Control) 7(3), June 2001 Supplement, pp1–10

Li Wenhua (2001) *Agro-Ecological Farming Systems in China*. Man and the Biosphere Series Volume 26. UNESCO, Paris

Lobao, L (1990) *Locality and Inequality: Farm and Industry Structure and Socio-Economic Conditions*. State University of New York Press, New York

Lobao, L M, Schulman, M D and Swanson, L E (1993) 'Still going: recent debate on the Goldschmidt Hypothesis', *Rural Sociology* 58(2), pp277–288

Lobstein, T, Millstone, E, Lang, T and van Zwanenberg, P (2001) *The lessons of Phillips. Questions the UK government should be asking in response to Lord Phillips' Inquiry into BSE*. Centre for Food Policy, Thames Valley University, London

Long, N and Long, A (1992) *Battlefields of Knowledge*. Routledge, London

Lopez, B (1986) *Arctic Dreams. Imagination and Desire in a Northern Landscape*. Harvill, London

Lopez, B (1998) *About this Life. Journeys on the Threshold of Memory*. Harvill, London

Losey, J E, Rayor, L S and Carter, M E (1999) 'Transgenic pollen harms monarch larvae', *Nature* 399, pp214

Mabey, R (1996) *Flora Britannica*. Sinclair Stevenson, London

MacRae, R J, Henning, J and Hill, S B (1993) 'Strategies to overcome barriers to the development of sustainable agriculture in Canada: the role of agribusiness', *Journal of Agricultural and Environmental Ethics* 6, pp21–51

MAFF (1999) *Reducing Farm Subsidies – Economic Adjustment in Rural Areas.* Working Paper 2. Economic and Statistics Group of MAFF, London

Maffi, L (1999) 'Linguistic diversity' in Posey, D (ed) (1999) *Cultural and Spiritual Values of Biodiversity*. IT Publications and UNEP, London

Malla, Y B (1997) 'Sustainable use of communal forests in Nepal', *Journal of World Forest Resource Management* 8, pp51–74

Mangan, J and Mangan, M S (1998) 'A comparison of two IPM training strategies in China: the importance of concepts of the rice ecosystem for sustainable pest management', *Agriculture and Human Values* 15, pp209–221

Manning, R E (1989) 'The nature of America: visions and revisions of wilderness', *Natural Resources Journal* 29, pp25–40

Mason, C F (1996) *Biology of Freshwater Pollution*. 3rd edition. Addison, Wesley Longman, Harlow

Mason, C F (1998) 'Habitats of the song thrush *Turdus philomelos* in a largely arable landscape', *Journal of Zoology, London* 244, pp89–93

Matteson, P C, Gallagher, K D and Kenmore, P E (1992) 'Extension of integrated pest management for planthoppers in Asian irrigated rice' in Denno, R F and Perfect, T J (eds) *Ecology and the Management of Planthoppers*. Chapman and Hall, London

Maturana, H R and Varela, F J (1992) *The Tree of Knowledge. The Biological Roots of Human Understanding*. Revised Edition. Shambhala, Boston and London

Maxwell, S and Frankenberger, T (1992) *Household food security: concepts, indicators, measurements*. IFAD, Rome

McBey, M A (1985) 'The therapeutic aspects of gardens and gardening: an aspect of total patient care', *Journal of Advanced Nursing* 10, pp591–595

McGinnis, M V (ed) (1999) *Bioregionalism*. Routledge, London and New York

McGloughlin, M (1999) 'Ten reasons why biotechnology will be important to the developing world', *AgBioForum* 2(3–4), pp163–174 (www.agbioforum.org/Default/mcgloughlin.htm)

McKean, M A (1985) 'The Japanese experience with scarcity: management of traditional common lands', *Environmental Reviews* 9(1), pp63–88

McKean, M A (1992) 'Management of traditional common lands (*Iriaichi*) in Japan' in Bromley, D W (ed) *Making the Commons Work*. Institute for Contemporary Studies Press, San Francisco, California

McNeely, J A and Scherr, S J (2001) *Common Ground, Common Future. How ecoagriculture can help feed the world and save wild biodiversity*. IUCN and Future Harvest, Geneva

McPartlan, H C and Dale, P J (1994) 'An assessment of gene transfer by pollen from field-grown transgenic potatoes to non-transgenic potatoes and related species', *Transgenic Research* 3, pp216–225

Mead, P S, Slutsker, L and Dietz, V (1999) 'Food related illness and death in the US', *Emerging Infectious Diseases* 5, pp607–625

Miles, G (1992) 'To hear an old voice' in Cronon, W, Miles, G and Gitlin, J (eds) (1992) *Under an Open Sky. Rethinking America's Western Past.* W W Norton and Co, New York, pp52–70

Millstone, E and van Zwanenberg, P (2001) 'Politics of expert advice: from the early history of the BSE saga', *Science and Public Policy* 28(2), pp99–112

Minami, K, Seino, H, Iwama, H and Nishio, M (1998) 'Agricultural land conservation'. OECD Workshop on Agri-Environmental Indicators. York, 22–25 September. OECD, Paris COM/AGR/CA./ENV/EPOC (98) 78

Minor, H C, Morris, C G, Mason, H L, Hanty, R W, Stafford, G K and Fritts, T G (1999) *Corn: 1999 Missouri Crop Performance.* University of Missouri-Columbia Special Report 521, University of Missouri, Columbia

Molina, F S (1998) 'The wilderness world is respected greatly: the Yoeme (Yaqui) truth from the Yoeme communities of Arizona and Sonora, Mexico' in Maffi, L (ed) *Language, Knowledge and the Environment.* Smithsonian Institution Press, Washington, DC

Monarch Butterfly Research Symposium (1999) 'Butterflies and *B.t.* corn pollen'. Chicago, 2 November 1999 (www.fooddialogue.com/monarch/newresearch.html)

Mooney, J D and Reiley, A C (1931) *Onward Industry.* Harper, New York

Moorehead, A (2000) *The Fatal Impact: An Account of the Invasion of the South Pacific.* Penguin Books, London

Muir, J (1911) *My First Summer in the Sierra.* Houghton Mifflin, Boston (reprinted in 1988 by Canongate Classics, Edinburgh)

Muir, J (1992) *The Eight Wilderness-Discovery Books.* Diadem Books, London and Seattle

Murphy, B (1999) *Cultivating Havana: Urban agriculture and food security in Cuba.* Food First Development Report 12. Food First, California

Myers, N (1998) 'Lifting the veil on perverse subsidies', *Nature* 392, pp327–328

Myers, N, Mittermeier, R A, Mittermeier, C G, de Fonseca, G A B and Kent, J (2000) 'Biodiversity hotspots for conservation priorities', *Nature* 403, pp853–858

Nabhan, G P and St Antoine, G (1993) 'The loss of floral and faunal story' in Kellert, S R and Wilson, E O (eds) *The Biophilia Hypothesis.* Island Press, Washington, DC

Naess, A (1992) 'Deep ecology and ultimate premises', *Society and Nature* 1(2), pp108–119

Narayan, D and Pritchett, L (1996) *Cents and Sociability: Household Income and Social Capital in Rural Tanzania.* Policy Research Working Paper 1796. The World Bank, Washington, DC

Nash, R (1973) *Wilderness and the American Mind.* Yale University Press, New Haven

National Audit Office (NAO) (1998) *BSE: The Cost of a Crisis.* NAO, London

National Landcare Programme (2001) Canberra, Australia (www.dpie.gov.au/landcare)

National Society of Allotments and Leisure Gardeners (www.nsalg.co.uk)

Newby, H (1988) *The Countryside in Question*. Hutchinson, London

Norse, D, Li Ji and Zhang Zheng (2000) *Environmental Costs of Rice Production in China: Lessons from Hunan and Hubei*. Aileen Press, Bethesda

NRC (2000) *Our Common Journey: Transition towards sustainability*. Board on Sustainable Development, Policy Division, National Research Council. National Academy Press, Washington, DC

Nuffield Council on Bioethics (1999) *Genetically Modified Crops: The Social and Ethical Issues*. Nuffield Council on Bioethics, London

O'Riordan, T (1999) *Dealing with scientific uncertainties*. Unpublished report. School of Environmental Sciences, University of East Anglia, Norwich

Oakerson, R J (1992) 'Analysing the commons: a framework' in Bromley D W (ed) (1992) *Making the Commons Work*. Institute for Contemporary Studies Press, San Francisco

Organization for Economic Cooperation and Development (OECD) (1997a) *Evaluating economic instruments for environmental policy*. OECD, Paris

OECD (1997b) *Helsinki Seminar on Environmental Benefits from Agriculture*. OECD/ GD(97)110, Paris

OECD (2000) *Environmental indicators for agriculture: methods and results. The stocktaking report – land conservation*. OECD, Paris COM/AGRI/CA./ENV/EPOC (99) 128/REVI

Oelschlaeger, M (1991) *The Idea of Wilderness*. Yale University Press, New Haven

Oerlemans, N, Proost, J and Rauwhost, J (1997) 'Farmers' study groups in the Netherlands' in van Veldhuizen, L, Waters-Bayer, A, Ramirez, R, Johnson, D A and Thompson, J (eds) *Farmers' Research in Practice*. IT Publications, London

Ofwat (1992–1998) *Annual returns from water companies – water compliances and expenditure reports*. Office of Water Services, Birmingham

Okri, B (1996) 'Joys of Story Telling' in *Birds of Heaven*. Penguin, Harmondsworth

Olson, M (1965) *The Logic of Collective Action: Public Goods and the Theory of Groups*. Harvard Press, London

Olson, M (1982) *The Rise and Decline of Nations: Economic Growth, Stagflation and Social Rigidities*. Yale University Press, New Haven

Oplinger, E S, Martinka, M J and Schmitz, K A (1999) 'Performance of transgenic soyabeans in the northern US'. University of Wisconsin (www. biotech-info.net/soyabean_performance.pdf)

Orr, D W (2000) 'Ideasclerosis: part two', *Conservation Biology* 14(6), pp1571–1573

Orr, D W (2002) *The Nature of Design*. Oxford University Press, Oxford

Oster, G, Thompson, D, Edelsberg, J, Bird, A P and Colditz, G A (1999) 'Lifetime health and economic benefits of weight loss among obese people', *American Journal of Public Health* 89, pp1536–1542

Ostrom, E (1990) *Governing the Commons: The Evolution of Institutions for Collective Action*. Cambridge University Press, New York

Ostrom, E (1998) *Social capital: a fad or fundamental concept?* Centre for the Study of Institutions, Population and Environmental Change, Indiana University, US

Pain, D J and Pienkowski, M W (eds) (1997) *Farming and Birds in Europe*. Academic Press Ltd, London

Palmer, I (1976) *The New Rice in Asia: Conclusions from Four Country Studies*. UNRISD, Geneva

Pannell, D (2001) 'Salinity policy: a tale of fallacies, misconceptions and hidden assumptions', *Agricultural Science* 14(1), pp35–37

Pearce, D and Tinch, R (1998) 'The true price of pesticides' in Vorley, W and Keeney, D (eds) *Bugs in the System*. Earthscan, London

Pearce, D W and Turner, R H (1990) *Economics of Natural Resources and the Environment*. Harvester Wheatsheaf, New York

Peiretti, R (2000) 'The evolution of the no till cropping system in Argentina'. Paper presented to Conference on the Impact of Globalisation and Information on the Rural Environment, 13–15 January, Harvard University, Cambridge, Massachusetts

Peluso, N L (1996) 'Reserving value: conservation ideology and state protection of resources' in DuPuis, E M and Vandergeest, P (eds) *Creating Countryside*. Temple University Press, Philadelphia

Perelman, M (1976) 'Efficiency in agriculture: the economics of energy' in Merril, R (ed) *Radical Agriculture*. Harper and Row, New York

Peter, G, Bell, M M, Jarnagin, S and Bauer, D (2000) 'Coming back across the fence: masculinity and the transition to sustainable agriculture', *Rural Sociology* 65(2), pp215–233

Petersen, C, Drinkwater, L E and Wagoner, P (2000) *The Rodale Institute's Farming Systems Trial: The First 15 Years*. Rodale Institute, Pennsylvania

Petersen, P, Tardin, J M and Marochi, F (2000) 'Participatory development of non-tillage systems without herbicides for family farming: the experience of the center-south region of Paraná', *Environmental Development and Sustainability* 1, pp235–252

Peterson, W L (1997) *Are Large Farms More Efficient?* Staff Paper P97-2. Department of Applied Economics, University of Minnesota, Minnesota

PHL (1999) 'Public Health Laboratory Service – facts and figures' (www.phls.co.uk/facts/)

Picardi, A C and Siefert, W W (1976) 'A tragedy of the commons in the Sahel', *Techology Review* 14, pp35–54

Pickett, J A (1999) 'Pest control that helps control weeds at the same time', *BBSRC Business*, April. Biotechnology and Biological Sciences Research Council (BBSRC), Swindon

Pimentel, D, Acguay, H, Biltonen, M, Rice, P, Silva, M, Nelson, J, Lipner, V, Giordano, S, Harowitz, A and D'Amore, M (1992) 'Environmental and economic cost of pesticide use', *Bioscience*, 42(10), pp750–760

Pimentel, D, Harvey, C, Resosudarmo, P, Sinclair, K, Kunz, D, McNair, M, Crist, S, Shpritz, L, Fitton, L, Saffouri, R and Blair, R (1995) 'Environmental and economic costs of soil erosion and conservation benefits', *Science* 267, pp1117–1123

Pingali, P L and Roger, P A (1995) *Impact of Pesticides on Farmers' Health and the Rice Environment*. Kluwer Academic Press, The Netherlands

Pinstrup-Andersen, P (1999) 'Developing appropriate policies', *IFPRI 2020 Vision Focus 2*, Brief 9 of 10. International Food Policy Research Institute, Washington, DC

Pinstrup-Andersen, P and Cohen, M (1999) 'World food needs and the challenge to sustainable agriculture'. Paper for Conference on Sustainable Agriculture: New Paradigms and Old Practices? Bellagio Conference Centre, Italy, 26–30 April 1999

Pinstrup-Andersen, P, Pandya-Lorch, R and Rosegrant, M (1999) *World Food Prospects: Critical Issues for the Early 21st Century*. IFPRI, Washington, DC

Pinto, Y M, Kok, P and Baulcombe, D C (1998) 'Genetically engineered resistance to RYMV disease is now a reality' in DFID (1998) Plant Science Research Programme. *Abstract from Annual Programme Report for 1997*. University of Wales, Bangor

Performance and Innovation Unit (PIU) (1999) *Rural Economies*. PIU, Cabinet Office, London

Platteau, J-P (1997) 'Mutual insurance as an elusive concept in traditional communities', *Journal of Development Studies* 33(6), pp764–796

Poffenberger, M and Zurbuchen, M S (1980) *The economics of village life in Bali.* The Ford Foundation, New Delhi

Poffenberger, M and McGean, B (eds) (1998) *Village Voices, Forest Choices.* Oxford University Press, New Delhi

Popkin, B (1998) 'The nutrition transition and its health implications in lower-income countries', *Public Health Nutrition* 1(1), pp5–21

Portes, A and Landolt, P (1996) 'The downside of social capital', *The American Prospect* 26, pp18–21

Posey, D (ed) (1999) *Cultural and Spiritual Values of Biodiversity.* IT Publications and UNEP, London

Postel, S and Carpenter, S (1997) 'Freshwater ecosystem services' in Daily, G (ed) (1997) *Nature's Services: Societal Dependence on Natural Ecosystems.* Island Press, Washington, DC

Potrykus, I (1999) 'Vitamin-A and iron-enriched rices may hold key to combating blindness and malnutrition: a biotechnology advance', *Nature Biotechnology* 17, p37

Potter, C (1998) *Against the Grain: Agri-Environmental Reform in the USA and European Union.* CAB International, Wallingford

Pretty, J N (1991) 'Farmers' extension practice and technology adaptation: Agricultural revolution in 17th–19th century Britain', *Agriculture and Human Values* VIII, pp132–148

Pretty, J N (1995a) *Regenerating Agriculture: Policies and Practice for Sustainability and Self-Reliance.* Earthscan, London; National Academy Press, Washington, DC; ActionAid, Bangalore

Pretty, J N (1995b) 'Participatory learning for sustainable agriculture', *World Development* 23(8), pp1247–1263

Pretty, J N (1998) *The Living Land: Agriculture, Food and Community Regeneration in Rural Europe.* Earthscan Publications Ltd, London

Pretty, J N (2000a) *Towards Sustainability in English Agriculture: The Multifunctional Role for Agriculture in the 21st Century.* The Countryside Agency, Cheltenham

Pretty, J N (2000b) 'Can sustainable agriculture feed Africa', *Environment, Development and Sustainability* 1, pp253–274

Pretty, J N (2001) 'The rapid emergence of genetically-modified crops in world agriculture', *Environmental Conservation* 28(3), pp248–262

Pretty, J N and Pimbert, M (1995) 'Beyond conservation ideology and the wilderness myth', *Natural Resources Forum* 19(1), pp5–14

Pretty, J N and Shah, P (1997) 'Making soil and water conservation sustainable: from coercion and control to partnerships and participation', *Land Degradation and Development*, 8, pp39–58

Pretty, J N and Frank, B (2000) 'Participation and Social Capital Formation in Natural Resource Management: Achievements and Lessons'. Plenary Paper for International Landcare 2000 Conference, Melbourne, Australia, 2–5 March 2000

Pretty, J N and Hine, R (2000) 'The promising spread of sustainable agriculture in Asia', *Natural Resources Forum (UN)* 24, pp107–121

Pretty, J N and Ball, A (2001) *Agricultural Influences on Emissions and Sequestration of Carbon and Emerging Trading Options.* CES Occasional Paper 2001-03, University of Essex, Colchester

Pretty, J N and Hine, R (2001) *Reducing Food Poverty with Sustainable Agriculture: A Summary of New Evidence.* Final Report from the SAFE-World Research Project, Feb 2001. University of Essex, Colchester

Pretty, J N and Ward, H (2001) 'Social capital and the environment', *World Development* 29(2), pp209–227

Pretty, J N and Buck, L (2002) 'Social capital and social learning in the process of natural resource management' in Barrett, C (ed) *Understanding Adoption Processes for Natural Resource Management Practices for Sustainable Agricultural Production in Sub-Sahara Africa.* CAB International, Wallingford

Pretty, J N, Thompson, J and Kiara, J K (1995) 'Agricultural regeneration in Kenya: the catchment approach to soil and water conservation', *Ambio* XXIV(1), pp7–15

Pretty, J N, Brett, C, Gee, D, Hine, R, Mason, C F, Morison, J I L, Raven, H, Rayment, M and van der, Bijl G (2000) 'An assessment of the total external costs of UK agriculture', *Agricultural Systems* 65(2), pp113–136

Pretty, J N, Brett, C, Gee, D, Hine, R, Mason, C, Morison, J, Rayment, M, van der Bijl, G and Dobbs, T (2001) 'Policy challenges and priorities for internalising the externalities of modern agriculture', *Journal of Environmental Planning and Management* 44(2), pp263–283

Pretty, J N, Mason, C F, Nedwell, D B and Hine, R E (2002) *A Preliminary Assessment of the Environmental Damage Costs of the Eutrophication of Freshwaters in England and Wales.* Report for the Environment Agency. University of Essex, Colchester

Pretty, J N, Morison, J I L and Hine, R E (2002) 'Reducing food poverty with sustainable agriculture', *Agriculture, Ecosystems and Environment*, in press, pp1–18

Prince, H (1988) 'Art and agrarian change, 1710–1815' in Cosgrove, D and Daniels, S (1988) *The Iconography of Landscape*. Cambridge University Press, London

Putnam, R D with Leonardi, R and Nanetti, R Y (1993) *Making Democracy Work: Civic Traditions in Modern Italy*. Princeton University Press, Princeton, New Jersey

Putnam, R (1995) 'Bowling alone: America's declining social capital', *Journal of Democracy* 6(1), pp65–78

Royal Academy of Arts (1981) *The Great Japan Exhibition. The Art of the Edo*. Royal Academy of Arts, London

Rackham, O (1986) *The History of the Countryside*. Dent, London

RAFI-USA (1998) *The Peanut Project. Farmer-focused Innovation for Sustainable Peanut Production*. RAFI, Pittsboro, North Carolina

Raju, G (1998) 'Institutional structures for community based conservation' in Kothari, A, Pathak, N, Anuradha, R V and Taneja, B (1998) *Communities and Conservation: Natural Resource Management in South and Central Asia*. Sage, New Delhi

Raybould, A F and Gray, A J (1993) 'GM crops and hybridisation with wild relatives: a UK perspective', *Journal of Applied Ecology* 130, pp199–219

Rayment, M, Bartram, H and Curtoys, J (1998) *Pesticide Taxes: A Discussion Paper*. Royal Society for the Protection of Birds, Sandy, Beds

RCEP (1996) *Sustainable Use of Soil*. 19th Report of the Royal Commission on Environmental Pollution. Cmnd 3165. HMSO, London

Read, M A and Bush, M N (1999) 'Control of weeds in GM sugar beet with glufosinate ammonium in the UK', *Aspects of Applied Biology* 52, pp401–406

Reader, J (1997) *Africa: A Biography of a Continent*. Hamish Hamilton, London

Rees, W E (1997) 'Why Urban Agriculture?' Paper for IDRC Development Forum on Communities Feeding People (www. Cityfarmer.org/rees.html)

Reicosky, D C, Dugas, W A and Torbert, H A (1997) 'Tillage-induced soil carbon dioxide loss from different cropping systems', *Soil and Tillage Research* 41, pp105–118

Reij, C (1996) 'Evolution et impacts des techniques de conservation des eaux et des sols'. Centre for Development Cooperation Services, Vrije Univeristeit, Amsterdam

Rengasamy, S, Devavaram, J, Prasad, R, Erskine, A, Balamurugan, P and High, C (2000) *The Land Without a Farmer Becomes Barren* (thaan vuzhu nilam thariso). SPEECH, Ezhil Nagar, Madurai, India

Repetto, R and Baliga, S S (1996) *Pesticides and the Immune System: The Public Health Risks*. WRI, Washington, DC

Ribaudo, M O, Horan, R D and Smith, M E (1999) *Economics of Water Quality Protection from Nonpoint Sources: Theory and Practice*. Agricultural Economic Report 782. Economic Research Service, US Department of Agriculture, Washington, DC

Rissler, J and Melon, M (1996) *The Ecological Risks of Engineered Crops*. MIT Press, Cambridge

Robins, N and Simms, A (2000) 'British aspirations', *Resurgence* 201 (July–August), pp6–9

Rola, A and Pingali, P (1993) *Pesticides, Rice Productivity and Farmers – An Economic Assessment.* IRRI, Manila and WRI, Washington, DC

Röling, N G and Wagemakers, M A E (eds) (1997) *Facilitating Sustainable Agriculture.* Cambridge University Press, Cambridge

Röling, N G (1997) 'The soft side of land: socio-economic sustainability of land use systems'. Invited Paper for Theme 3, Proceedings of Conference on Geo-Information for Sustainable Land Management, ITC, Enschede, 17–21 August

Röling, N G (2000) 'Gateway to the global garden. Beta/gamma science for dealing with ecological rationality'. Eighth Annual Hopper Lecture, 24 October, University of Guelph, Canada

Röling, N G and Jiggins, J (1997) 'The ecological knowledge system', in Röling, N R and Wagemakers, M A (eds) *Social Learning for Sustainable Agriculture.* Cambridge University Press, Cambridge

Rolston, H (1997) 'Nature is for real: is nature a social construct?' in Chappell, T (ed) *Philosophy of the Environment.* Edinbugh University Press, Edinburgh

Rominger, R (2000) Speech at opening of USDA's fifth farmers' markets season by Deputy Secretary Rich Rominger, June 9 (www.ams.usda.gov/farmers markets/romigerfmkt1.htm)

Rosegrant, M W, Leach, N and Gerpacio, R (1997) 'Alternative futures for world cereal and meat consumption', *Proceedings of Nutrition Science* 58, pp219–234

Rosset, P (1997) 'Alternative agriculture and crisis in Cuba', *IEEE Technology and Society Magazine* 16(2), pp19–26

Rosset, P (1998) 'Alternative agriculture works: the case of Cuba', *Monthly Review* 50(3), pp137–146

Rosset, P (1999) *The Multiple Functions and Benefits of Small Farm Agriculture.* Food First Policy Brief No 4. Food First/Institute for Food and Development Policy, Oakland, California

Rowley, J (1999) *Working with social capital.* Report for Department for International Development, London

Roy, S D and Jackson, P (1993) 'Mayhem in Manas: the threats to India's wildlife reserves' in Kemf, E (ed) *Indigenous Peoples and Protected Areas.* Earthscan, London

Royal Society (1998) *Genetically Modified Plants for Food Use.* Royal Society, London

Royal Society, US National Academy of Sciences, Brazilian Academy of Sciences, Chinese Academy of Sciences, Indian National Academy of Sciences, Mexican Academy of Sciences, and Third World Academy of Sciences (2000) *Transgenic Plants and World Agriculture.* Royal Society, London

Ryszkowski, C (1995) 'Managing ecosystem services in agricultural landscapes', *Nature and Resources* 31(4), pp27–36

Samson, C (2002) *Innu Naskapi-Montagnais. A Way of Life that Does Not Exist. Canada and the 'Extinguishment of the Innu'.* ISER Press, St Johns, Newfoundland

Sanchez, P A (2000) 'Linking climate change research with food security and poverty reduction in the tropics', *Agriculture, Ecosystems and the Environment* 82, pp371–383

Sanchez, P A and Jama, B A (2000) *Soil fertility replenishment takes off in East and Southern Africa.* ICRAF, Nairobi

Sanchez, P A, Buresh, R J and Leakey, R R B (1999) 'Trees, soils and food security', *Philosophical Transactions of the Royal Society of London* B 253, pp949–961

SARE (1998) *Ten Years of SARE.* CSREES, USDA, Washington, DC (www.sare.org)

Sarin, M (2001) 'Disempowerment in the name of participatory forestry? Village forests management in Uttarakhand', *Forest, Trees and People Newsletter* 44, pp26–34

Saxena, D, Flores, S and Stotzky, G (1999) 'Insecticidal toxin in root exudates from *Bt* corn', *Nature* 401, p480

Schama, S (1996) *Landscape and Memory.* Fontana Press, London

Schwarz, W and Schwarz, D (1999) *Living Lightly.* Green Books, Bideford

Scoones, I (1994) *Living with Uncertainty: New Directions in Pastoral Development in Africa.* IT Publications, London

Scoones, I (1998) *Sustainable Rural Livelihoods: A Framework for Analysis.* IDS Discussion Paper, 72, University of Sussex, Falmer

Scott, J (1998) *Seeing Like a State. How Certain Schemes to Improve the Human Condition Have Failed.* Yale University Press, New Haven

Scruton, R (1998) 'Conserving the past' in Barnett, A and Scruton, R (eds) *Town and Country.* Jonathan Cape, London

Segundad, P (1999) 'Malaysia – Kadazan' in Posey, D (ed) (1999) *Cultural and Spiritual Values of Biodiversity.* IT Publications and UNEP, London

Seidl, A (2000) 'Economic issues and the diet and the distribution of environmental impact', *Ecological Economics* 34(1), pp5–8

Selman, P (1993) 'Landscape ecology and countryside planning: vision, theory and practice', *Journal of Rural Studies* 9(1), pp1–21

Sessions, G (ed) (1995) *Deep Ecology for the 21st Century.* Shambhala Publications, Boston

Shah, A (1998) 'Participatory process of organising effective community-based groups'. International Workshop on Community-Based Natural Resource Management. World Bank, Washington DC, 10–14 May 1998

Shanin, T (1986) *The Roots of Otherness. Russia's Turn of the Century.* Macmillan, London

Shankar, D (1998) 'Conserving a community resource: medicinal plants' in Kothari, A, Pathak, N, Anuradha, R V and Taneja, B (eds) *Communities and Conservation: Natural Resource Management in South and Central Asia.* Sage, New Delhi

Shelley, M (1818) *Frankenstein, or The Modern Prometheus.* Reprinted in 1963, J M Dent & Sons Ltd, London

Short, C (2000) 'Common land and ELMS: a need for policy innovation in England and Wales', *Land Use Policy* 17, pp121–133

Short, C and Winter, M (1999) 'The problem of common land: towards stakeholder governance', *Journal of Environmental Planning and Management* 42(5), pp613–630

Shrestha, B (1998) 'Involving local communities in conservation: the case of Nepal' in Kothari, A, Pathak, N, Anuradha, R V and Taneja, B (1998)

*Communities and Conservation: Natural Resource Management in South and Central Asia.*
Sage, New Delhi

Shrestha, K B (1997) 'Community forestry: policy, legislation and rules'. Paper
presented at national workshop on Community Forestry and Rural Development, 24–26 July, Lalipur, Nepal

Singh, K and Bhattacharya, S (1996) 'The salt miners' cooperatives in the Little
Rann of Kachchh in Gujarat' in Singh, K and Ballabh, V (1997) *Cooperative
Management of Natural Resources.* Sage, New Delhi

Singh, K and Ballabh, V (1997) *Cooperative Management of Natural Resources.* Sage,
New Delhi

Siriwardena, G M, Ballie, S R, Buckland, G T, Fewster, R M, Marchant, J H and
Wilson, J D (1998) 'Trends in the abundance of farmland birds: a quantitative
comparison of smoothed Common Birds Census indices', *Journal of Applied
Ecology* 35, pp24–43

Smaling, E M A, Nandwa, S M and Janssen, B H (1997) 'Soil fertility in Africa
is at stake' in Buresh, R J, Sanchez, P A and Calhoun, F (eds) *Replenishing Soil
Fertility in Africa.* Soil Science Society of America Publication No 51. SSSA,
Madison, Wisconsin

Small Farm Viability Project (1977) *The Family Farm in California.* Governor's
Office of Planning and Research, Sacramento, California.

Smit, J, Ratta, A and Nasr, J (1996) *Urban Agriculture: Food Jobs and Sustainable Cities.*
UNDP, New York

Smith, S and Piacentino, D (1996) *Environmental taxation and fiscal reform: analysis
of implementation issues.* Final Report EV5V-CT894-0370, DGXI, Brussels

Smith, B (1985) *European Vision and the South Pacific.* Yale University Press, New
Haven and London (2nd edition)

Smith, B (1987) *Imagining the Pacific.* Yale University Press, New Haven

Smith, L C and Haddad, L (1999) *Explaining Child Malnutrition in Developing
Countries: A cross-country analysis.* Research Report 111 (March 2000). IFPRI,
Washington, DC

Socorro Castro, A R (2001) 'Cienfuegos, the capital of urban agriculture in
Cuba' (www.cityfarmer.org/cubacastro.html)

Soil Association (2000) *The Organic Food and Farming Report.* Soil Association, Bristol

Sorrenson, W J, Duarte, C and Portillo, J L (1998) *Economics of no-till compared to
conventional systems on small farms in Paraguay.* Soil Conservation Project MAG-
GTZ, Eschborn, Germany

Soulé, M E and Lease, G (eds) (1995) *Reinventing Nature? Responses to Postmodern
Deconstruction.* Island Press, Washington, DC

Spinage, C (1998) 'Social change and conservation misrepresentation in Africa',
*Oryx* 32(3), pp1–12

SPWD (1998) *Joint Forest Management Update.* Society for the Promotion of
Wastelands Development, New Delhi

Steinberg, T (1995) *Slide Mountain. Or the Folly of Owning Nature.* University of
California Press, Berkeley

Steinbeck, J (1939) *The Grapes of Wrath.* Minerva, London (republished in 1990)

Steiner, R, McLaughlin, L, Faeth, P and Janke, R (1995) 'Incorporating externality costs in productivity measures: a case study using US agriculture' in Barbett, V, Payne, R and Steiner, R (eds) *Agricultural Sustainability: Environmental and Statistical Considerations.* John Wiley, New York, pp209–230

Stewart, L, Hanley, N and Simpson, I (1997) *Economic Valuation of the Agri-Environment Schemes in the UK.* Report to HM Treasury and the Ministry of Agriculture, Fisheries and Food. Environmental Economics Group, University of Stirling, Stirling

Stoll, S and O'Riordan, T (2002) *Protecting the Protected: Managing Biodiversity for Sustainability.* Cambridge University Press, Cambridge

Stott, P (1999) *The organic myth.* School of Oriental and African Studies, University of London, London

Stren, M and Alton, E W F W (1998) 'Gene therapy for cystic fibrosis', *Biologist* 45(1), pp37–40

Suffolk Horse Society, Woodbridge, Suffolk (www.suffolkhorsesociety. org.uk)

Suzuki, D (1999) 'Finding a new story' in Posey, D (ed) (1999) *Cultural and Spiritual Values of Biodiversity.* IT Publications and UNEP, London

Suzuki, D and Oiwa, K (1996) *The Japan We Never Knew.* Allen and Unwin, Tokyo and London

Swift, M J, Vandermeer, J, Ramakrishnan, P G, Anderson, J M, Ong, C K and Hawkins, B A (1996) 'Biodiversity and agroecosystem function' in Mooney, H A, Cushman, J H, Medina, E, Sala, O E and Schulze, E D (eds) *Functional Roles of Biodiversity: A Global Perspective.* SCOPE, John Wiley and Sons, Cheltenham

Swingland, I, Bettelheim, E, Grace, J, Prance, G and Saunders, L (eds) (2002) 'Carbon, Biodiversity, Conservation and Income', *Transactions of the Royal Society (Series A: Mathematical, Physical and Engineering Sciences),* in press

Swiss Agency for Environment, Forests and Landscape and Federal Office of Agriculture (2000) *Swiss agriculture on its way to sustainability.* SAEFL and FOA, Basel

Swiss Agency for Environment, Forests and Landscape (1999) *The Environment in Switzerland: Agriculture, Forestry, Fisheries and Hunting* (www.admin.ch/buwal/ e/themen/partner/landwirt/ek21u00.pdf)

Tall, D (1996) 'Dwelling; making peace with space and place' in Vitek, W and Jackson, W (eds) *Rooted in the Land: Essays on Community and Place.* Yale University Press, New Haven and London

Tanksley, S D and McCouch, S R (1997) 'Seed banks and molecular maps: unlocking genetic potential from the wild', *Science* 277, pp1063–1066

Taylor, M (1982) *Community, Anarchy and Liberty.* Cambridge University Press, Cambridge

Temple, R K G (1986) *China: Land of Discovery and Invention.* Patrick Stephenson, Wellingborough

*The Lancet* (1999) 'Editorial', 253(9167), p1811

Thompson, E P (1975) *Whigs and Hunters.* Penguin, Harmondsworth

Thoreau, H D (1837–1853) *The Writings of H D Thoreau*, Volumes 1–6 (published 1981–2000). Princeton University Press, Princeton, New Jersey

Thoreau, H D (1902) *Walden or Life in the Woods*. Henry Frowde, Oxford University Press, London, New York and Toronto

Thoreau, H D (1906) *The Writings of H D Thoreau* (including 'A Winter Walk'). Houghton Mifflin, Boston

Thoreau, H D (1987) *Maine Woods*. Harper and Row, New York

Thrupp, L A (1996) *Partnerships for Sustainable Agriculture*. World Resources Institute, Washington, DC

Tibble, A (ed) *The Journal; Essays; The Journey from Essex*. Carcanet New Press, Manchester

Tilman, D (2000) 'Causes, consequences and ethics of biodiversity', *Nature* 405, pp208–211

Tilman, D (1998) 'The greening of the green revolution', *Nature* 396, pp211–212

Tinker, P B (2000) *Shades of Green: A Review of UK Farming Systems*. RASE, Stoneleigh Park, Warwickshire

Tönnies, F (1887) *Gemeinschaft und Gesellschaft (Community and Association)*. Routledge and Kegan Paul, London (1955 edition)

Turner, F J (1920) *The Frontier in American History*. Holt, Rinehart and Winston, New York

UNEP/WCMC (2001) *United Nations List of Protected Areas*. UNEP World Conservation Monitoring Centre (www.wcmc.org.uk/)

United Nations Population Fund (1999) *World Population Prospects – The 1998 Revision*. United Nations, New York

Uphoff, N (1992) *Learning from Gal Oya: Possibilities for Participatory Development and Post-Newtonian Science*. Cornell University Press, Ithaca, New York

Uphoff, N (1993) 'Grassroots organisations and NGO in rural development: opportunity with diminishing stakes and expanding markets', *World Development* 21(4), pp607–622

Uphoff, N (1998) 'Understanding social capital: learning from the analysis and experience of participation' in Dasgupta, P and Serageldin, I (eds). *Social Capital: A Multiperspective Approach*. World Bank, Washington, DC

Uphoff, N (1999) 'What can be learned from SRI in Madagascar about meeting future food needs'. Paper for conference: Sustainable Agriculture New paradigms and Old Practices? Bellagio Conference Centre, Italy, 26–30 April

Uphoff, N (2000) 'Agroecological implications of the system of rice intensification (SRI) in Madagascar', *Environmental Development and Sustainability* 1(3–4), pp297–313

Uphoff, N (ed) (2002) *Agroecological Innovations*. Earthscan, London

US Senate Science Committee (2000) *Seeds of Opportunity: An Assessment of the Benefits, Safety and Oversight of Plant Genomics and Agricultural Biotechnology*. 106th Congress, 2nd Session, Committee Print 106-B, 13 April, Washington, DC

US Department of Agriculture (USDA) (1999) National Agricultural Statistics Service (NASS). Data on planted acres of major crops (www.usda.gov/nass/)

USDA (2000) *Factbook*. USDA, Washington, DC (www.usda.gov)

USDA (1998) *A Time to Act. National Commission on Small Farms.* USDA, Washington, DC

USDA (2001a) 'Farm size and numbers data'. USDA, Washington, DC (www. usda.gov)

USDA (2001b) 'Farm statistics'. USDA, Washington, DC (www.ers.usda. gov/statefacts)

van de Fliert, E (1997) 'From pest control to ecosystem management: how IPM training can help?' Paper presented to International Conference on Ecological Agriculture, Indian Ecological Society, Chandigarh, India, November 1997

van der Bijl, G and Bleumink, J A (1997) *Naar een Milieubalans van de Agrarische Sector* [Towards an Environmental Balance of the Agricultural Sector]. Centre for Agriculture and Environment (CLM), Utrecht

van der Ploeg, R R, Ehlers, W and Sieker, F (1999) 'Floods and other possible adverse effects of meadowland area decline in former West Germany', *Naturwissenschaften* 86, pp313–319

van der Ploeg, R R, Hemsmeyer, D and Bachmann, J (2000) 'Postwar changes in landuse in former West Germany and the increasing number of inland floods' in Marsalek et al (eds) *Flood Issues in Contemporary Water Management.* Kluwer Academic Publications, The Netherlands, pp115–123

van Veldhuizen, L, Waters-Bayer, A, Ramirez, R, Johnson, D A and Thompson, J (eds) (1997) *Farmers' Research in Practice.* IT Publications, London

van Weperen, W , Proost, J and Röling, N G (1995) 'Integrated arable farming in the Netherlands' in Röling, N G and Wagemakers, M A E (eds) (1997) *Facilitating Sustainable Agriculture.* Cambridge University Press, Cambridge

Vandergeest, P and DuPuis, E M (1996) 'Introduction' in DuPuis, E M and Vandergeest, P (eds) *Creating Countryside: The Politics of Rural and Environmental Discourse.* Temple University Press, Philadelphia

von der Weid, J M (2000) *Scaling up and Scaling Further Up.* AS-PTA, Rio de Janeiro

Vorley, W and Keeney, D (eds) (1998) *Bugs in the System: Redesigning the Pesticide Industry for Sustainable Agriculture.* Earthscan, London

Waibel, H and Fleischer, G (1998) *Kosten und Nutzen des chemischen Pflanzenschutzes in der Deutsen Landwirtschaft aus Gesamtwirtschaftlicher Sicht.* Vauk-Verlag, Kiel

Wall, P G, de Louvais, J, Gilbert, R J and Rowe, B (1996) 'Food poisoning: notifications, laboratory reports and outbreaks – where do the statistics come from and what do they mean?' *Communicable Disease Report* 6 (7), R94-100

Waltner-Toews, D and Lang, T (2000) 'A new conceptual base for food and agricultural policy', *Global Change and Human Health* 1(2), pp2–16

Ward, H (1998) 'State, association, and community in a sustainable democratic polity: towards a green associationalism' in Coenen, F, Huitema, D and O'Toole, L J (eds) (1998) *Participation and the Quality of Environmental Decision Making.* Kluwer Academic Publishers, Dordrecht

Weida, W J (2000) *A Citizen's Guide to the Regional Economic and Environmental Effects of Large Concentrated Dairy Operations.* The Colorado College, Colorado Springs, CO, and The Global Resource Action Center for the Environment (GRACE), Factory Farm Project (www.factoryfarm.org)

Weissman, J (ed) (1995) *City Farmers: Tales from the Field*. GreenThumb, New York

Weissman, J (ed) (1995) *Tales from the Field. Stories by GreenThumb Gardeners*. GreenThumb, New York

Wesselink, W (2001) '240 ha of beef cattle', *Agrifuture – Europa Agribusiness Magazine*, Spring I, pp16–18

West, P C and Brechin, S R (1992) *Resident People and National Parks*. University of Tucson Press, Arizona

Wheeler, E B, Wiley, K N and Winne, M (1997) 'A tale of two systems' in *In Context* 42, pp25–27

White T (2000) 'Diet and the distribution of environmental impact', *Ecological Economics* 34(1), pp145–153

Whitlock, R (1979) *In Search of Lost Gods. A Guide to British Folklore*. Phaidon Press, Oxford

World Health Organization (WHO) (1998) *Obesity. Preventing and managing the global epidemic*. WHO Technical Report 894. WHO, Geneva

WHO (2001) *Food and Health in Europe. A Basis for Action*. Regional Office for Europe, WHO, Copenhagen

WHO Regional Office for Europe (2000) *Urban Agriculture in St Petersburg, Russian Federation*. WHO, Copenhagen, Denmark

Willers, B (2000) 'A response to "current normative concepts in conservation" by Callicott et al', *Conservation Biology* 14(2), pp570–572

Willis, K, Garrod, G and Saunders, C (1993) *Valuation of the South Downs and Somerset Levels Environmentally Sensitive Areas*. Centre for Rural Economy, University of Newcastle upon Tyne

Wilson, E O (1988) 'The current state of biological diversity' in Wilson, E O and Peter, F M (eds) *Biodiversity*. National Academy Press, Washington, DC

Winrock International (2000) *Biotechnology and Global Development Challenges*. Little Rock, Arkansas

Wise, R, Hart, T, Cars, O, Streulens, M, Helmuth, R, Huovinen, P and Sprenger, M (1998) 'Antimicrobial resistance', *British Medical Journal* 317, pp609–610

Wolfe, M (2000) 'Crop strength through diversity', *Nature* 406, pp681–682

Wood, S, Sebastien, K and Scherr, S J (2000) *Pilot Analysis of Global Ecosystems*. IFPRI and WRI, Washington, DC

Woolcock, M (1998) 'Social capital and economic development: towards a theoretical synthesis and policy framework', *Theory and Society* 27, pp151–208

Wordie, J R (1983) 'The chronology of English enclosure 1500–1914', *Economic History Review* 36, pp483–505

World Bank (1995) *Pakistan: The AKRSP. A Third Evaluation*. Report No 15157, Washington, DC

World Bank/FAO (1996) *Recapitalisation of Soil Productivity in Sub-Saharan Africa*. World Bank and FAO, Washington, DC, and Rome

World Neighbors (WN) (1999) *After Mitch: Towards a Sustainable Recovery in Central America*. World Neighbors, Oklahoma City, US

Worster, D (1993) *The Wealth of Nature: Environmental History and the Ecological Imagination*. Oxford University Press, New York

WPPR (1997) *Annual Report of the Working party on Pesticide Residues: 1997.* MAFF, London

Ya, T (1999) 'Factors Influencing Farmers' Adoption of Soil Conservation Programme in the Hindu Kush Himalayan (HKH) Region'. Paper for Issues and Options in the Design of Soil and Water Conservation Projects: A Workshop, University of Wales, Bangor/University of East Anglia, Llandudno, Conwy, UK, 1–3 February 1999

Young, J, Humphreys, M, Abberton, M, Robbins, M and Webb, J (1999) *The risks associated with the introduction of GM forage grasses and forage legumes.* Report for MAFF (RG0219) research project. Institute of Grassland and Environmental Research, Aberystwyth

Zhu, Y, Chen, H, Fen, J, Wang, Y, Li, Y, Zhen, J, Fan, J, Yang, S, Hu, L, Leaung, H, Meng, T W , Teng, A S, Wang, Z and Mundt, C C (2000) 'Genetic diversity and disease control in rice', *Nature* 406, pp718–722

# Index

'A great balance between storytelling and analysis which points to the critical need for gaining control over resources'
**Jacqueline Ashby, CIAT, Colombia**

'A vision- and thought-provoking book which puts the reader at ease'
**Andy Ball, University of Essex, UK**

'I certainly think it bears out the desire to blend the story telling with the theory. The stories bring a very human dimension to what can be a dry area'
**David Beckingsale, Department of Natural Resources and Environment, Victoria, Australia**

'Terrific – very important. An extremely interesting and stimulating book'
**Ted Benton, University of Essex, UK**

'An absorbing book with an excellent writing style, full of good argument and supported by evidence. I like the broad reach and the coupling of the developed and the developing world, particularly in the context of local knowledge, the commons [and] the connection of consumers to producers'
**Phil Bradley, University of Hull, UK**

'Very good indeed. It manages to bridge academia and more general writing very well. It's timely, innovative, and the watercolours are a delight'
**Lynda Brown, foodwriter, London, UK**

'Thought-provoking and readable, with interesting, sudden changes in the landscapes and locations under discussion'
**Edward Cross, Abbey Farm, Norfolk, UK**

'A book that you can read straight through rather than a reference book to look up what you want to know. And a book that's about land and community needs stories – I think the balance is great'
**Jan Deane, Northwood Farm, Devon, UK**

'Inspiring with a clear sense of re-connectedness'
**Amadou Diop, Kutztown, Pennsylvania, US**

'An excellent and very readable book'
**Thomas Dobbs, South Dakota State University, US**

'A very interesting and timely read, written with some passion'
**David Favis-Mortlock, Queen's University Belfast, Northern Ireland**

'A seminal work akin to Rachel Carson's effort in the 60s, with a good balance between story telling and critical analysis'
**Bruce Frank, University of Queensland, Australia**

'Beautifully written. The implications of the book's ideas are deep and extensive'
**Julia Guivant, University of Florianopolis, Brazil**

'What makes this book more readable and interesting than the typical writing on sustainable agriculture is the endless examples from around the world viewed through the eyes of the people who do the work on the ground, told through their voices, and experienced through their frustrations. An excellent primer on our food system'
**Brian Halweil, Worldwatch Institute, US**

'I love the use of stories, and the descriptions of their importance. How we tell our story shapes our actual behaviour on the earth and with one another'
**Hal Hamilton, Sustainability Institute, Vermont, US**

'Should be required reading for anyone involved in agriculture around the globe'
**Justin Hardy, Department of Agriculture – Western Australia, Australia**

'Most convincing – a tremendous amount of work has gone into the documentation of the argument'
**John Landers, Associacao de Plantio Direto no Cerrado, Brazil**

'A fabulous book'
**Tim Lang, Thames Valley University, UK**

'An original book, and full of good ideas'
**Howard Lee, Imperial College at Wye, UK**

'Satisfying as a conceptual whole – a resounding message to those responsible for Welsh farming to build a better future around their traditional distinctive cultural and spiritual relationship with the land'
**David Lort-Phillips, Lawrenny farm, Wales**

'An excellent analysis of all the problems and potential solutions in relation to sustainable agriculture'
**Simon Lyster, The Wildlife Trusts, UK**

'Conveys the commonality of issues and themes (and vulnerabilities) that are evident in the rural sector – the diversity and convergence of values, circumstances and practices and frustrations which define the rural landscape and its occupants'
**Joe Morris, Cranfield University, UK**

'*Refreshingly fluent narrative, brimming full of stories and metaphors*'
Tim O'Riordan, University of East Anglia, UK

'*This book will be a great asset*'
Roberto Peiretti, No-Till Farmer's Association, Argentina

'*The ideas flow well, with a clear and direct prose. The illustrations are evocative and relevant*'
Michel Pimbert, IIED, UK

'*Agri-Culture* is definitely going to shape the thinking processes of many in the years to come'
S Rengasamy, SPEECH, Madurai, India

'*A wonderful book, put together with such vision and passion*'
Mark Ritchie, Institute of Agriculture and Trade Policy, US

'*The writing is elegantly crafted and refreshingly trans-disciplinary in outlook*'
Colin Samson, University of Essex, UK

'*A very interesting, readable and compelling book*'
Sara Scherr, Forest Trends, Washington, DC, US

'*Full of supporting evidence and clear arguments*'
Norman Uphoff, Cornell University, US

'*An accessible but in-depth discussion of many of the crucial issues, and it should be influential*'
Hugh Ward, University of Essex, UK

'*A clear sense of pace in the text. The story lines move along strongly*'
Drennan Watson, Aberdeen, Scotland

'*Chock-filled with interesting and pertinent information presented in a most engaging way*'
Jane Weissman, New York City, US

'*Telling stories and making it personal, when combined with some scholarship and analysis of the contemporary scene is a particularly persuasive approach. It will be a great addition to the growing body of recent work on the relation between man, land, nature and agriculture*'
Mark Winne, Hartford Food System, US

'*A superb volume. This is a valuable monograph that all policy-makers, scholars and farmers must read to understand their roles and responsibilities*'
Vo-Tong Xuan, Angiang University, Vietnam